Nonlinear Analysis and Continuum Mechanics

Papers for the 65th Birthday of James Serrin

Springer
New York
Berlin
Heidelberg
Barcelona
Budapest
Hong Kong
London
Milan
Paris
Santa Clara
Singapore
Tokyo

Giuseppe Buttazzo Giovanni Paolo Galdi
Ermanno Lanconelli Patrizia Pucci
Editors

Nonlinear Analysis and Continuum Mechanics

Papers for the 65th Birthday of James Serrin

Springer

Giuseppe Buttazzo
Department of Mathematics
University of Pisa
Pisa, I-56127, Italy

Giovanni Paolo Galdi
Department of Engineering
University of Ferrara
Ferrara, I-44100, Italy

Ermanno Lanconelli
Department of Mathematics
University of Bologna
Bologna, I-40126, Italy

Patrizia Pucci
Department of Mathematics
University of Perugia
Perugia, I-06123, Italy

With 3 Figures

Library of Congress Cataloging-in-Publication Data
Nonlinear analysis and continuum mechanics : papers for the 65th birthday of James Serrin / [editors] Giovanni Paolo Galdi . . . [et al.].
p. cm.
Includes index.
ISBN-13:978-1-4612-7455-1 e-ISBN-13:978-1-4612-2196-8
DOI: 10.1007/978-1-4612-2196-8

1. Mathematical analysis. 2. Nonlinear theories. 3. Continuum mechanics. I. Serrin, J. (James), 1926- . II. Galdi, Giovanni P. (Giovanni Paolo), 1947-
QA300.5.N663 1997
531'.01'515355—dc21 97-26098

Printed on acid-free paper.

Production managed by Karina Mikhli; manufacturing supervised by Johanna Tschebull.
Typeset by BookComp, Inc., Grand Rapids, MI using TeX.

9 8 7 6 5 4 3 2 1

ISBN-13:978-1-4612-7455-1 Springer-Verlag New York Berlin Heidelberg SPIN 10633994

Contents

Foreword

A conference, to celebrate the 65th birthday of Professor James Serrin, was held at the University of Ferrara from 26 to 30 October 1992, on the occasion of the sixth centenary of the foundation of the university.

The dual purpose of this conference was to honor James Serrin and to bring together mathematicians in four main areas of Professor Serrin's research interests: partial differential equations, variational methods, fluid mechanics, and thermodynamics. These fields, with their significant historical connotation, still remain the object of active research and development, and possess a remarkable degree of interrelation in their pure and applied aspects. Thus the organizing committee, consisting of G. Buttazzo, G.P. Galdi, E. Lanconelli, and P. Pucci, had the opportunity to invite a number of well-known mathematicians and friends of James Serrin to make this meeting a success.

An honorary committee was also formed, including Professor A. Rossi, Rector of the University of Ferrara, Professor C. Bighi, President of the Accademia Scienze di Ferrara, Professor G. Del Piero, Decano of the Faculty of Engineering of the University of Ferrara, and Professor L. Zanghirati, Director of the Department of Mathematics of the University of Ferrara. The conference was further supported by G.N.A.F.A., G.N.F.M., and the Mathematics Committee of the Consiglio Nazionale delle Ricerche, together with the University of Ferrara and the Cassa di Risparmio of Ferrara, to all of whom the organizing committee expresses their most sincere thanks.

Many of the lectures delivered at the conference are collected in the present volume, together with several further papers by invited speakers who were unable to attend. The editors cordially thank all these colleagues for their contributions. In addition to the mathematical papers appearing here, there is also included an exposition by P. Pucci concerning aspects of the life of James Serrin and his mathematical work.

The second morning of the conference was dedicated to the memory of Professor Lamberto Cesari, a close friend of James Serrin. For this occasion the late Professor Gaetano Fichera delivered a tribute to Lamberto Cesari's remarkable contributions to mathematical analysis.

A significant moment of the conference occurred when Professor Serrin was granted the Laurea honoris causa in Ingegneria dei Materiali by the University of Ferrara, presented by the Rector, Professor Antonio Rossi. The Lectio Doctoralis of James Serrin is also included among the contributions.

In conclusion, the editors thank Dr. Giovanni Leoni and Dr. Enzo Vitillaro for their valuable assistance in the preparation of this volume.

The Editors

Contributors

Antonio Ambrosetti
Scuola Normale Superiore, Piazza dei Cavalieri, 7-56126 Pisa, Italy

Gianfranco Capriz
Dipartimento di Matematica, Universitá di Pisa, Via Buonarroti, 2-56127 Pisa, Italy

David E. Edmunds
School of Mathematics, University of Sussex, Falmer, Brighton BN1 9QH, Sussex, United Kingdom

Mariano Giaquinta
Dipartimento di Matematica Applicata, Universitá di Firenze, Via S. Marta, 3 50139 Firenze, Italy,

Robin J. Knops
Department of Mathematics, Heriot-Watt University, Edinburgh, EH14 4AS, Scotland

Olga Ladyzhenskaya
Saint Petersburg Branch of Steklov Institute, Fontanka 27, Saint Petersburg, Russia 191011

C. Lupoli
Department of Mathematics, Heriot-Watt University, Edinburgh, EH14 4AS, Scotland

Giacomo Mazzini
Dipartimento di Matematica, Universitá di Pisa, Via Buonarroti, 2 56127 Pisa, Italy

J.Bryce McLeod
Department of Mathematics, University of Pittsburgh, Pittsburgh, PA 15260, USA

Giuseppe Modica
Dipartimento di Matematica Applicata, Universitá di Firenze, Via S. Marta, 3 50139 Firenze, Italy

Lambertus A. Peletier
Department of Mathematics, University of Leiden, P.O. Box 9512, 2300R.A. Leiden, The Netherlands

Patrizia Pucci
Dipartimento di Matematica, Universitá di Perugia, Via Vanvitelli 1, Perugia, I-06123, Italy

James Serrin
Department of Mathematics, University of Minnesota, Minneapolis, MN 55455, USA

Vsevolod A. Solonnikov
Saint Petersburg Branch of Steklov Institute, Fontanka 27, Saint Petersburg, Russia 191011

Jiří Souček
Akademie věd České republiky, Matematický Ústav, Zitná, 25, 11567 Praha, ČR

Kazunaga Tanaka
Department of Mathemetics, School of Science, Nagoya University, Chikusa-ku, Nagoya 464, Japan

1
An Appreciation of James Serrin

Patrizia Pucci

I am particularly happy to have the opportunity of these proceedings to write a profile of one of my distinguished teachers, with whom I have also had the good fortune to collaborate for so many years, Professor James Serrin. Before entering into my recollection of these years of close association, I would first desire to mention the tribute that Professor Clifford Truesdell wrote in the volume of papers, Analysis and Continuum Mechanics, dedicated to James Serrin on the occasion of his sixtieth birthday, published by Springer-Verlag, in gratitude for his many years of work as coeditor of the *Archive for Rational Mechanics and Analysis*.

The outstanding contributions which Professor Serrin has made, and still makes, to the field of mathematical analysis and its applications are well known. Many of his mathematical discoveries are now in the form of basic principles, whose importance should remain undiminished far into the future. His name is attached to several theorems and methods, and many of his early results have become classical and are well represented in modern textbooks on partial differential equations, calculus of variations, real analysis, fluid mechanics, and thermodynamics: the Gilbarg–Serrin theorem on isolated singularities, the Serrin symmetry theorem, the Meyers-Serrin theorem, the work on quasi-linear elliptic equations, the Alexandrov-Serrin moving plane technique, the Serrin integral of the calculus of variations, the Peletier-Serrin uniqueness theorem, and the Serrin version of the second law of thermodynamics may be quoted here, as well as other more recent contributions.

From this partial list of his most famous results we can also see a further quality of his thought: the aptitude in collaborating with his colleagues and in generously sharing his ideas with them. This aspect of his personality was particularly impressed on me at the conference organized in Minneapolis in May 1991 to celebrate his sixty-fifth birthday, at which a distinguished group of colleagues spoke about his mathematical career.

Another distinction of his temperament is the elegance of his taste. This is evident in much of his magisterial work, in which each of his theorems has the beauty of an essential form, as in the artistic masterpieces of my country which he loves so much. This is a further reason why his work has influenced others so much, and why many of his theorems have opened new directions in mathematical research.

Professor Serrin was born in Chicago on November 1, 1926. He first lived in Evanston, where he studied at Evanston Township High School. He then attended

Northwestern University for two years, majoring in Electrical Engineering. After receiving the Bachelor of Arts degree from Western Michigan College, he then went to Indiana University, where in 1951 he received his doctorate with a thesis on the hydrodynamical theory of cavitation, written under the supervision of Professor *David Gilbarg*. Following his initial appointment as Fine Instructor of Mathematics at Princeton University in 1951, he became C.L.E. Moore Instructor of Mathematics at the Massachusetts Institute of Technology in the period 1952–1954. In Boston he met and married Barbara West in September 1952, the marvellous and patient partner of his life, who gave him the warmth of a lovely family, gladdened by the birth of three beautiful daughters: Martha, Elizabeth, and Janet. She has shared with him all the anxieties of life, as well as the recognition which his work receives today.

He joined the faculty at the University of Minnesota, as Assistant Professor in Mathematics until 1956, as Associate Professor until 1959, and finally as Professor from then until now. At Minnesota he was named head of the School of Mathematics in the period 1964–1965, and was appointed Regents' Professor of Mathematics in 1969.

Professor Serrin has received many honors for his work. He was awarded the degree *laurea honoris causa* by the University of Sussex in 1972, and by the Universities of Ferrara and Padua in 1992. From the American Mathematical Society he received the George David Birkhoff Award in Applied Mathematics in 1973, and from the Indiana University their Distinguished Alumni Award in 1979. He was elected as a member of the National Academy of Sciences in 1980 and of the American Academy of Arts and Sciences in 1984, and as a fellow of the American Academy for the Advancement of Science in 1980. To commemorate his sixtieth birthday, a series of papers dedicated to him in the *Archive for Rational Mechanics and Analysis* were collected in the volume *Analysis and Continuum Mechanics* (Springer-Verlag, 1989), whose dedicatory preface by Professor Clifford Truesdell was already mentioned above. Finally, two conferences were held to celebrate his sixty-fifth birthday: at the University of Minnesota in 1991 and at a corresponding meeting organized by the University of Ferrara in 1992, from which the present volume developed.

He has served on numerous professional committees, most recently being Chair of the American Mathematical Society Committee for Progress in Mathematics. He was president of the Society for Natural Philosophy in 1969–1970 and on the council of the American Mathematical Society in 1972–1974. From 1969 to 1986 he was a coeditor of the *Archive for Rational Mechanics and Analysis*, where his dedicated editorial efforts continued and enhanced its reputation among the great journals of mathematical analysis. He has been on the Editorial Board of the above journal since 1963, and also the *Journal of Differential Equations* since 1976, the *Communications in Partial Differential Equations* since 1978, the *Rendiconti del Circolo Matematico di Palermo* since 1980, the *Atti del Seminario Matematico e Fisico dell'Universita di Modena* since 1985, *Asymptotic Analysis* since 1988, and *Differential and Integral Equations* since 1990.

Professor Serrin has been a visiting professor at Stanford University, the University of Chicago, The Johns Hopkins University, the University of Sussex, the Mittag–Leffler Institute, the Universities of Naples, Perugia, and Modena, Oxford University, and the National Science Council in Taiwan. Furthermore, his list of invited addresses, symposia, and departmental colloquia is almost endless. I recall here his superb lectures at Cortona and Milan in 1981, and at Varenna in 1982, which I had the good fortune to attend, and which inspired some of our mutual research: an aspect of my research life for which I am deeply grateful. As further recognition of his achievements in mathematical research, Professor Serrin delivered invited addresses at the International Congresses of Mathematicians in Nice, 1970, and in Warsaw, 1983, and has lectured at numerous major scientific conferences.

During his teaching career, James Serrin directed the research of twelve mathematicians who obtained, under his supervision their doctoral degrees in pure and applied mathematics. He is more than cordial with aspiring young people and his office door is always open to all who desire to discuss problems with him. His generosity goes even further, since Barbara and James share their marvellous house with many colleagues: young and old, unknown or famous.

Professor Serrin's publications comprise more than 160 serious papers, whose range is best appreciated from the following complete list. He has written extensive and important papers on partial and ordinary differential equations, on the calculus of variations, on real analysis, on fluid mechanics, on the constitutive theory of materials, and on the modern foundations of phenomenological thermodynamics. His fundamental contributions to these fields were the basis for the award of the laurea honoris causa in Padua on the four-hundredth anniversary celebration of the association of Galileo with the University of Padua in 1992.

I conclude with a review of some of Professor Serrin's main research, without specifically mentioning his many collaborators, some of whom were already quoted at the beginning of this appreciation.

Partial Differential Equations

In general elliptic theory Professor Serrin had two early fundamental papers [13], [25] covering the Harnack inequality and exterior Dirichlet problems. Papers [14] and [46] are fundamental to the problem of isolated singularities and had great impact on the future work in singularity theory. The extensive monograph [41] contains basic a priori estimates for quasi-linear equations, following the celebrated method of De Giorgi, Moser, and Nash. Two related papers [52], [53] deal with the corresponding behavior of solutions of parabolic quasi-linear equations. A slightly earlier result [44] shows the care which is required in defining weak solutions of elliptic equations with continuous coefficients, while [68] contains a general maximum principle for solutions of quasi-linear divergence structure differential inequalities. Concerning the solvability of boundary value problems for quasi-linear equations, in [56] he discovered the negative mean curvature boundary condition for the minimal surface equation, while in [59] there are elegant conditions for the existence of solutions of the infinite data problem. Finally [64], which has been extensively quoted and is the basis for two chapters of the celebrated book on elliptic equations by Gilbarg and Trudinger, is a general treatise on the Dirichlet problem for quasi-linear differential equations with several independent variables. The main contributions in symmetry theory are in [65] and [73]: the first contains one of the nicest proofs by Professor Serrin, the second initiated the use of the moving plane technique for the study of partial differential equations—the so-called Alexandrov—Serrin method is now a standard tool, yielding the most subtle symmetry results.

In another series of papers [74], [77], [82], [85], [87] he applied the maximum principle to obtain a remarkable group of Liouville theorems and gradient estimates for quasi-linear elliptic equations. Here I quote from [77] the simplest of these results:

Let u be an entire solution of the equation $\Delta u + f(Du) = 0$. *Suppose that f is continuous and that both u and Du are bounded. Then* $u \equiv$ *constant.*

From his recent work in semilinear and quasi-linear equations I mention here [95], [108], [150]; [117], [118]; [122], [153] concerning uniqueness, existence, and nonexistence theorems for the canonical equationì $\Delta u + f(u) = 0$; multivalued radial solutions of prescribed mean curvature equations; and the extension of the Critical Dimension Theorem of Brezis and Nirenberg to the case of polyharmonic operators. A series of papers [149], [154], on hyperbolic equations may also be noticed.

Ordinary Differential Equations

In the 1990s Professor Serrin took up the study of the asymptotic stability of solutions of nonautonomous second-order equations and systems, extending earlier and celebrated work on the subject due to Levin and Nohel, to Artstein and Infante, and to a number of their followers, see [128], [129], [133], [134], and [136]. Finally, a recent paper [139] may be mentioned, concerning the asymptotic integration of quasi-linear differential systems, a problem that had been initiated by the work of Weyl and Wintner for linear equations.

Calculus of Variations

Serrin's main contributions on the subject are contained in [28], [101], and [112]. The early paper on the definition of variational integrals started a great deal of Itȧlian work and has been extensively quoted. The second is from a series of papers on the Mountain Pass theory, detailing the structure of the critical set arising in such mini-max constructions. The third is a fundamentally useful identity for solutions of variational problems, related to the invariance ideas of Noether. Another early paper [40] introduced the idea of variational integrals for functions whose ìderivatives are measures.

Real Analysis

The celebrated papers [27] and [42] can be found in every introductory theory on Sobolev spaces and are still quoted nowadays. The first shows that we can find a representative of a weakly differentiable function which has the usual partial derivatives almost everywhere, and also contains a remarkable example of a strongly differentiable function which does not possess a differential at any point (the fakir's bed of nails). The second paper proves the fundamental fact that weak and strong differentiation are identical. The chain rule for differentiation is taken up in [62], for functions which have derivatives only almost everywhere, leading to the following general form of the change of variables rule for Lebesgue integration:

Suppose that g has a finite derivative almost everywhere on $[a, b]$ and that f is integrable on $[c, d]$. Then $(f \circ g)g'$ is integrable, and the change of variables formula

$$\int_{g(\alpha)}^{g(\beta)} f(x)\,dx = \int_{\alpha}^{\beta} f(g(s))g'(s)_s\,ds$$

holds for all α, β in the domain of g, if and only if the left-hand integral is an absolutely continuous function of β.

I should also note the simple proof of the Lebesgue Set Theorem [61], which is included in Walter Rudin's treatise *Real and Complex Analysis.*

Fluid Mechanics

The existence theorems in [1] and [3] for hydrodynamical free boundary problems and the qualitative results in [6] and [8] are his main contributions in free boundary theory. In [6], among other things, the shape of a symmetric obstacle with least wake drag in the Helmoltz–Kirchhoff theory was determined; in fact, it is essentially this shape which is now used for modern semitrucks to improve performance efficiency. The papers [18]–[20] and [32] in Navier–Stokes theory provide the foundation for the modern energy method of hydrodynamic stability, and the interior regularity of solutions of the Navier–Stokes equations. Reference [34] is a well-known paper on the status of the initial value problem in its classical period. In [55] he established the asymptotic behavior of solutions of the laminar boundary layer equations, thus providing a theoretical underpinning for similarity solutions of Blasius type. The dynamical structure of tornados is discussed in [76], in an extensive mathematical and physical study of flow patterns at the point of impact of tornados. The papers [21] and [106] in constitutive theory are a monument in rational mechanics and remain classical sources. Finally, I must mention the fundamental article [17] on fluid mechanics, one of the major post-war treatises in this field, which has served as a handbook for countless students of the subject.

Thermodynamics

The main papers [80], [88], [90], [93], and [107] on the laws of thermodynamics for irreversible processes are based on a thorough familiarity with the works of Kelvin,

Clausius, Gibbs, and Carathéodory. On the same subject may also be mentioned four papers [81], [104], [130], [151] on the foundations of mechanics.

Besides the mathematical papers, there are three expository works which deserve mention: the contribution [84] to the American Mathematical Society volume on Hilbert Problems, the essay [99] on Applied Mathematics, dedicated to Lamberto Cesari, and the polemical paper [125] on the nature of thermodynamics. Two other introductory essays worth noting occur in the treatiseì [64] and in the paper of tornados [76].

I wish to close by saying that collaborating with Professor Serrin is a marvellous experience, not only for the great pleasure in viewing the elegance and beauty of his way of thinking of mathematics, but also, when possible, to share with him and Barbara the moments of relaxation in which his highly literate life appear: his interest in classical music and his sophisticated and ample collection of rare records; his sincere love of art and architecture and his splendid paintings (many by Professor Serrin's brother, Richard), and his selection of fine objects collected with Barbara year by year; his warm and large library with impressive and almost unique books on art; and the collection of photographs taken essentially during his scientific travels. From all these interests, which he shares day by day with his devoted wife Barbara, is evident the goodness of a life never trivial or middling.

James Serrin's publications

[1] Free boundaries and jets in the theory of cavitation (with D. Gilbarg). *J. Math. and Phys.*, **29** (1950), 1–12.

[2] Uniqueness theorems for two free boundary problems. *Amer. J. Math.*, **74** (1952), 492–506.

[3] Existence theorems for some hydrodynamical free boundary problems. *J. Rational Mech. Anal.*, **1** (1952), 1–48.

[4] Two hydrodynamical comparison theorems. *J. Rational Mech. Anal.*, **1** (1952), 563–572.

[5] Notes on hydrodynamics. Photoprinted by Princeton University, 1952, 218 pages.

[6] On plane and axially symmetric free boundary problems. *J. Rational Mech. Anal.*, **2** (1953), 563–575.

[7] A note on the wave equation. *Proc. Amer. Math. Soc.*, **5** (1954), 307–308.

[8] Comparison theorem for subsonic flows. *J. Math. and Physics*, **23** (1954), 27–45.

[9] On the Phragmen–Lindelőf theorem for elliptic partial differential equations. *J. Rational Mech. Anal.*, **3** (1954), 395–413.

[10] A uniqueness theorem for the parabolic equation $u_t = a(x)u_{xx} + b(x)u_x + c(x)u$, *Bull. Amer. Math. Soc.*, **60** (1954), 344.

[11] Uniqueness of axially symmetric subsonic flow past a finite body (with D. Gilbarg). *J. Rational Mech. Anal.*, **4** (1955), 169–175.

[12] A characterization of regular boundary points for second-order linear differential equations. *Bull. Amer. Math. Soc.*, **61** (1955), 224.

[13] On the Harnack inequality for linear elliptic equations. *J. Analyse Math.*, **4** (1956), 292–308.

[14] On isolated singularities of solutions of second-order linear elliptic equations (with D. Gilbarg). *J. Analyse Math.*, **4** (1956), 309–340.

[15] A note on harmonic functions defined in a half-plane. *Duke Math. J.*, **24** (1956), 523–526.

[16] On the Hölder continuity of quasi-conformal and elliptic mappings (with R. Finn). *Trans. Amer. Math. Soc.*, **89** (1958), 1–15.

[17] *Mathematical Principles of Classical Fluid Mechanics* (Monograph). Handbuch der Physik, **VIII/1** (1959), pp. 125–263. Russian translation: Foreign Literature Publishing House, Moscow, 1963, 265 pages.

[18] On the stability of viscous fluid motion. *Arch. Rational Mech. Anal.*, **3** (1959), 1–13.

[19] A note on the existence of periodic solutions of the Navier–Stokes equations. *Arch. Rational Mech. Anal.*, **3** (1959), 120–122.

[20] On the uniqueness of compressible fluid motions. *Arch. Rational Mech. Anal.*, **3** (1959), 271–288.

[21] On the derivation of stress-deformation relations for a Stokesian fluid. *J. Math. Mech.*, **8** (1959), 459–470.

[22] Poiseuille and Couette flow of non-Newtonian fluids. *Z. Angli. Math. Mech.*, **39** (1959), 295–299.

[23] On a fundamental theorem of the calculus of variations. *Acta Math.*, **102** (1959), 1–22.

[24] A new definition of the integral for non-parametric problems in the calculus of variations. *Acta Math.*, **102** (1959), 23–32.

[25] The exterior Dirichlet problem for second-order elliptic equations (with N. Meyers). *J. Math. Mech.*, **9** (1960), 513–538.

[26] On the area of curved surfaces. *Amer. Math. Monthly*, **68** (1961), 435–440.

[27] On the differentiability of functions of several variables. *Arch. Rational Mech. Anal.*, **7** (1961), 359–372.

[28] On the definition and properties of certain variational integrals. *Trans. Amer. Math. Soc.*, **101** (1961), 139–167.

[29] On the entropy change through a shock layer (with Y.C. Whang). *J. Aerospace Sci.*, **28** (1961), 990–991.

[30] *Dirichlet's Principle in the Calculus of Variations*. Proc. Symposia in Pure Math., vol. 4. American Mathematical Society, Providence, RI, 1961, pp. 17–22.

[31] Interior estimates for solutions of the Navier–Stokes equations. In *Partial Differential Equations and Continuum Mechanics*, (R. Langer, ed.). University of Wisconsin Press, 1961, pp. 376–378.

[32] On the interior regularity of weak solutions of the Navier–Stokes equations. *Arch. Rational Mech. Anal.*, **9** (1962), 187–195.

[33] Strong convergence in a product space. *Proc. Amer. Math. Soc.*, **13** (1962), 651–655.

[34] The initial value problem for the Navier–Stokes equations. In *Nonlinear Problems* (R.E. Langer, ed.). University of Wisconsin Press, 1963, pp. 69–98.

[35] Variational problems of minimal surface type, I (with H. Jenkins). *Arch. Rational Mech. Anal.*, **12** (1963), 185–212. See also entries [50], [59].

[36] A Harnack inequality for non-linear equations. *Bull. Amer. Math. Soc.*, **69** (1963), 481–486.

[37] Comparison and averaging methods in mathematical physics. In *Proprietà di Media e Teoremi di Confronto in Fisica Matematica.* Centro Internazionale Matematico Estivo, Rome, Edizioni Cremonese, 1965, pp. 1–87.

[38] A priori estimates for solutions of the minimal surface equation. *Arch. Rational Mech. Anal.*, **14** (1963), 376–383. See also entry [57].

[39] Mathematical Aspects of Boundary Layer Theory. Notes taken by H.K. Wilson, Department of Mathematics, University of Minnesota, 1963 (131 pages, multiplied typescript).

[40] Sublinear functions of measures and variational integrals (with C. Goffman). *Duke Math. J.*, **31** (1964), 159–178.

[41] Local behavior of solutions of quasi-linear equations. *Acta Math.*, **111** (1964), 247–302.

[42] $H = W$ (with Norman Meyers). *Proc. Nat. Acad. Sci.*, **51** (1964), 1055–1056.

[43] Removable singularities of solutions of elliptic equations. *Arch. Rational Mech. Anal.*, **17** (1964), 67–78. See also entry [48].

[44] Pathological solutions of elliptic differential equations. *Ann. Scuola Norm. Sup. Pisa, Sci. Fis. Math.*, **18** (1964), 385–387.

[45] Singularities of Solutions of Nonlinear Equations. *Proc. Symposia in Pure Math.*, vol. **17**, pp. 68–88. American Mathematical Society, Providence, RI, 1965.

[46] Isolated singularities of solutions of quasi-linear equations. *Acta Math.*, **113** (1965), 219–240.

[47] Theory of differentiation. Notes taken by T. Hatcher, Department of Mathematics, University of Minnesota, 1965 (135 pages, multiplied typescript).

[48] Removable singularities of solutions of elliptic differential equations, II. *Arch. Rational Mech. Anal.*, **20** (1965), 163–169.

[49] The Dirichlet problem for the minimal surface equation with infinite data (with H. Jenkins). *Bull. Amer. Math. Soc.*, **72** (1966), 102–106. See also entry [59].

[50] Variational problems of minimal surface type, II: Boundary value problems for the minimal surface equation (with H. Jenkins). *Arch. Rational Mech. Anal.*, **21** (1966), 321–342.

[51] Isolated singularities of solutions of linear elliptic equations (with H. Weinberger). *Amer. J. Math.*, **88** (1966), 258–272.

[52] Local behavior of solutions of quasi-linear parabolic equations (with D.G. Aronson). *Arch. Rational Mech. Anal.*, **25** (1967), 81–122.

[53] A maximum principle for nonlinear parabolic equations (with D.G. Aronson), *Ann. Scuola Norm. Sup. Pisa, Sci. Fis. Mat.*, **21** (1967), 291–305.

[54] The Dirichlet problem for quasi-linear elliptic equations with many independent variables. *Proc. Nat. Acad. Science*, **58** (1967), 1829–1835. See also entry [64].

[55] On the asymptotic behavior of velocity profiles in the Prandtl boundary layer theory, *Proc. Roy. Soc. London Ser. A*, **299** (1967), 491–507.

[56] The Dirichlet problem for the minimal surface equation in higher dimensions (with H. Jenkins). *J. Reine Angli. Math.*, **223** (1968), 170–187.

[57] Addendum to: A piori estimates for solutions of the minimal surface equation. *Arch. Rational Mech. Anal.*, **28** (1968), 149–154.

[58] On the mathematical basis of Prandtl's boundary layer theory: An example. *Arch. Rational Mech. Anal.*, **28** (1968), 217–225.

[59] Variational problems of minimal surface type, III: The Dirichlet problem with infinite data (with H. Jenkins). *Arch. Rational Mech. Anal.*, **29** (1968), 304–322.

[60] The behavior of similar solutions in a compressible boundary layer (with J.B. McLeod). *J. Fluid Mech.*, **34** (1968), 337–342.

[61] A new proof in differentiation theory. *Notices Amer. Math. Soc.*, **15** (1968), 1036.

Abstract; see also W. Rudin, *Real and Complex Analysis*, 2nd edn. McGraw-Hill, New York, 1974, pp. xii, 435, and 162–167.

[62] A general chain rule for derivatives and the change of variable formula for the Lebesgue integral (with D.E. Varberg). *Amer. Math. Monthly*, **76** (1969), 514–520.

[63] The existence of similar solutions for some laminar boundary layer problems (with J.B. McLeod). *Arch. Rational Mech. Anal.*, **31** (1969), 288–303.

[64] The problem of Dirichlet for quasi-linear elliptic differential equations with many independent variables. *Philos. Trans. Roy. Soc. London Ser., A*, **264** (1969), 413–496.

[65] On surfaces of constant mean curvature which span a given space curve. *Math. Z.*, **88** (1969), 77–88.

[66] Existence theorems for some compressible boundary layer problems. In *Qualitative Theory of Nonlinear Differential and Integral Equations. SIAM Stud. Appl. Math.*, vol. **5** 1970, pp. 35–42.

[67] The Dirichlet problem for surfaces of constant mean curvature. *Proc. London Math. Soc.*, **21** (1970), pp. 361–384.

[68] On the strong maximum principle for nonlinear second-order differential inequalities. *J. Funct. Anal.*, **5** (1970), 184–193.

[69] Boundary curvatures and the solvability of Dirichlet's Problem. *Proc. International Congress of Mathematicians* (Nice, 1970), vol. 2, Paris, 1970, 867–875.

[70] Curvature inequalities for surfaces over a disk (with H.F. Weinberger). In *Some problems of Mathematics and Mechanics—M.A. Lavrentieff Anniversary Volume.* Nauka Leningrad, 1970, pp. 242–250. English version: Amer. Math. Soc. Translation, vol. **104**, 1976, 223–231.

[71] Recent developments in the mathematical aspects of boundary layer theory. *Internat. J. Engng. Sci.*, **9** (1971), 233-240.

[72] Uniqueness and comparison theorems for nonlinear elliptic equations in divergence form (with J. Douglas, Jr. and T. Dupont). *Arch. Rational Mech. Anal.*, **42** (1971), 157–168.

[73] A symmetry problem in potential theory. *Arch. Rational Mech. Anal.*, **43** (1971), 304-318.

[74] Gradient estimates for solutions of nonlinear elliptic and parabolic equations. In *Contributions to Nonlinear Functional Analysis* (E. Zarantonello, ed.). University of Wisconsin Press, 1971, 565–601.

[75] Nonlinear elliptic equations of second order. Lectures at *Symposium on Partial Differential Equations*, Berkeley, 1971. Mimeographed notes (57 pages, multiplied typescript).

[76] The swirling vortex. *Philos. Trans. Roy. Soc. London Ser. A*, **271** (1972), 325–360.

[77] Entire solutions of nonlinear Poisson equations. *Proc. London Math. Soc.*, **24** (1972), 348–366.

[78] A note on the preceding paper of Amann. *Arch. Rational Mech. Anal.*, **44** (1972), 182–186.

[79] Rectilinear steady flow of simple fluids (with R.L. Fosdick). *Proc. Royal Soc. London*, **332** (1973), 311–333.

[80] Global properties of continuum thermodynamic processes (with R.L. Fosdick). *Arch. Rational Mech. Anal.*, **59** (1975), 97–109.

[81] On the axioms of classical mechanics. Department of Mathematics, University of Minnesota, 1974 (74 pages, multiplied typescript).

[82] Liouville theorems for nonlinear Poisson equations. In *Convegno Internazionale sui*

Metodi Valutativi nella Fisica–Matematica, Accad. Naz. Lincei, Problemi Attuali di Scienza e di Cultura, vol. **217**, 1975, pp. 207–215.

[83] Doomsday: On target? *Science Magazine*, **189**, July 11, 1975, pp. 86–87.

[84] The solvability of boundary value problems: Hilbert's twentieth problem. In *Mathematical Developments Arising from Hilbert Problems*, Proc. Sympos. Pure Math., vol. **28**, American Mathematical Society, Providence, RI, 1976, pp. 507–524.

[85] Liouville theorems and gradient bounds for quasi-linear elliptic systems. *Arch. Rational Mech. Anal.*, **66** (1977), 295–310.

[86] Optimal shapes for brittle beams under torsion (with H.F. Weinberger). In *Complex Analysis and its Applications—Jubilee Volume for I.Y. Vekua*, Nauka, Moscow, 1978, pp. 88–91.

[87] Gradient estimates and Liouville theorems for quasi-linear elliptic equations (with L.A. Peletier). *Ann. Scuola Norm. Sup. Pisa*, **5**, Ser. IV (1978), 65–104.

[88] The concepts of thermodynamics. In *Continuum Mechanics and Partial Differential Equations* (G.M. de la Penha et al., eds.). North-Holland, Amsterdam, 1978, pp. 411–451.

[89] On the impossibility of linear Cauchy and Piola–Kirchhoff constitutive theories for stress in solids (with R.L. Fosdick). *J. Elasticity*, **9** (1979), 83–89.

[90] Conceptual analysis of the classical second laws of thermodynamics. *Arch. Rational Mech. Anal.*, **70** (1979), 254–272.

[91] Foundations of thermodynamics. Lecture Notes, University of Naples, 1979 (150 pages, multiplied typescript).

[92] Phase transitions and interfacial layers for van der Waals fluids. In *Recent Methods in Nonlinear Analysis and Applications* (Proc. Fourth International Meeting of SAFA), (A. Canfora et al., eds.). Liguori Editore, Naples, 1981, pp. 169–175.

[93] The second law of thermodynamics for systems with approximate cycles (with B.D. Coleman and D.R. Owen). *Arch. Rational Mech. Anal.*, **77** (1981), 103–142.

[94] Uniqueness of solutions of semilinear Poisson equations (with K. McLeod). *Proc. Nat. Acad. Sci. USA*, **78** (1981), 6592–6598. See also entry [114].

[95] Uniqueness of positive solutions of semilinear equations in $\mathbb{R}^n$ (with L.A. Peletier), Archive Rational Mech. Anal., **81** (1983), 181-197. See also entry [110].

[96] The form of interfacial surfaces in Korteweg's theory of phase equilibria. *Quart. Appl. Math.*, **41** (1983), 357–364.

[97] The mechanical theory of fluid interfaces and Maxwell's rule (with E. Aifantis). *J. Coll. Interface Sci.*, **96** (1983), 519–547.

[98] The structure and laws of thermodynamics. *Proc. International Congress of Mathematicians* (Warsaw 1983), 1717–1728.

[99] Applied mathematics and scientific thought. In *Nonlinear Analysis and Optimization*. Lecture Notes in Mathematics, vol. **1107**, Springer-Verlag, New York, 1984, pp. 19–27.

[100] One-dimensional shock layers in Korteweg fluids (with R. Hagan). In *Phase Transformations and Material Instabilities in Solids.* Academic Press, New York, 1984, 113–128.

[101] Extensions of the mountain pass theorem (with P. Pucci). *J. Funct. Anal.*, **59** (1984), 185–210.

[102] A mountain pass theorem (with P. Pucci). *J. Differential Equations*, **60** (1985), 142–149.

[103] Esistenza ed unicità degli stati fondamentali per equazioni ellittiche quasilineari (with

B. Franchi and E. Lanconelli). *Rend. Accad. Naz. Lincei, Serie 8*, **79** (1985), 121–126. See also entries [119], [150].

[104] A continuum model for chemical mixture dynamics. In *Developments in Mechanics* (Proc. Midwestern Mechanics Conference), vol. **13**, 1985, pp. 59–64.

[105] Non-existence theorems for quasi-linear partial differential equations (with W.-M. Ni). *Suppl. Rend. Circolo Mat. Palermo*, **8** (1985), 171–185.

[106] On the thermomechanics of interstitial working (with J.E. Dunn). *Arch. Rational Mech. Anal.*, **88** (1985), 95–133.

[107] An outline of thermodynamical structure. In *New Perspectives in Thermodynamics*, (J. Serrin, ed.). Springer-Verlag, New York, 1986, 3–32.

[108] Existence and non–existence theorems for ground states of quasi-linear partial differential equations. The anomalous case (with W.-M. Ni). *Accad. Naz. Lincei, Atti dei Convegni*, **77** (1986), 231–257.

[109] Dynamic changes of phase in a van der Waals fluids (with R. Hagan). In *New Perspectives in Thermodynamics*. (J. Serrin, ed.). Springer-Verlag, New York, 1986, pp. 241–260.

[110] Uniqueness of non-negative solutions of semilinear equations in $\mathbb{R}^n$ (with L.A. Peletier). *J. Differential Equations*, **61** (1986), 380–397.

[111] Non-existence of singular solutions of quasi-linear partial differential equations (with W.-M. Ni). *Comm. Pure Appl. Math.*, **39** (1986), 379–399.

[112] A general variational identity (with P. Pucci). *Indiana Univ. Math. J.*, **35** (1986), 681–703.

[113] The laws of thermodynamics. *Rev. Mechanica*, **21** (1987), 361–379.

[114] Uniqueness of positive radial solutions of $\Delta u + f(u) = 0$ in $\mathbb{R}^n$ (with K. McLeod). *Arch. Rational Mech. Anal.*, **99** (1987), 115–145.

[115] The structure of the critical set in the mountain pass theory (with P. Pucci). *Trans. Amer. Math. Soc.*, **229** (1987), 115–132.

[116] Ground states for a prescribed mean curvature equation (with L.A. Peletier). *Proc. Amer. Math. Soc.*, **100** (1987), 694–700.

[117] Positive solutions of a prescribed mean curvature equation. In *Calculus of Variations and Partial Differential Equations* (S. Hildebrandt et al., eds.). Lecture Notes in Mathematics, vol. **1340**. Springer-Verlag, New York, 1988, pp. 248–255.

[118] Ground states for the prescribed mean curvature equation: The supercritical case (with F.V. Atkinson and L.A. Peletier). In *Nonlinear Diffusion Equations and Their Equilibrium States (I)* (W.-M. Ni et al., eds.). MSRI Publ., vol. **12**. Springer-Verlag, New York, 1988, pp. 51–74.

[119] Existence and uniqueness of ground states of quasilinear elliptic equations (with B. Franchi and E. Lanconelli). In *Nonlinear Diffusion Equations and Their Equilibrium States (I)*, MSRI Publ., vol. **12**. Springer-Verlag, New York, 1988, pp. 293–300. See also entry [150].

[120] Remarks on the first eigenspace for polyharmonic operators (with P. Pucci). *Atti Sem. Mat. Fis. Univ. Modena*, **36** (1988), 107–118.

[121] Asymptotic properties for solutions of strongly nonlinear ordinary differential equations (with P. Pucci). In *Conference on Differential Equations and Geometry. Rend. Sem. Mat. Univ. Politec. Torino*, Fascicolo Speciale, 1989, pp. 121–129.

[122] Critical exponents and critical dimensions for the polyharmonic operator (with P. Pucci). *J. Math. Pures Appl.*, **69** (1990), 55–83.

[123] Continuation and limit properties for solutions of strongly nonlinear second order differential equations (with P. Pucci). *Asymptotic Anal.*, **4** (1991), 97–160.

[124] Bounds for vertical points of solutions of prescribed mean curvature type equations, II (with F.V. Atkinson and L.A. Peletier). *Asymptotic Anal.*, **5** (1992), 283–310.

[125] The nature of thermodynamics. *Atti Sem. Mat. Fis. Univ. Modena*, **39** (1991), 445–472. Italian translation: Sulla natura della termodinamica. *Atti e Memorie Accad. Naz. Scienze Lettere e Arti di Modena, Ser. VII*, vol. **7** 1992, pp. 185–202.

[126] Global asymptotic stability for strongly nonlinear second order systems (with P. Pucci). In *Nonlinear Diffusion Equations and Their Equilibrium States (III)*. (N.G. Lloyd et al., eds.). Birkhäuser, Boston, 1992, pp. 437–449.

[127] Existence and non-existence for ground states of quasilinear elliptic equations (with H. Zou), *Arch. Rational Mech. Anal.*, **121** (1992), 101–130.

[128] Precise damping conditions for global asymptotic stability for nonlinear second-order systems (with P. Pucci). *Acta Math.*, **170** (1993), 275–307. See also entry [133].

[129] Continuation and limit behavior for damped quasi-variational systems (with P. Pucci). In *Degenerate Diffusions*, (W.-M. Ni et al., eds.). IMA Vols. in Math. Appl., vol. **47**. Springer-Verlag, New York, 1993, pp. 157–174.

[130] The equations of continuum mechanics as a consequence of group invariance. In *Advances in Continuum Mechanics*, (G. Ferrarese, ed.). Pitagora Editrice, Bologna, 1993, pp. 217–225.

[131] Asymptotic behavior of solutions of a nonstandard second-order differential equation (with W.A. Harris and P. Pucci). *Differential Integral Equations*, **6** (1993), 1201–1215.

[132] Existence and non-existence results for ground states of degenerate Laplace equations (with H. Zou). In *Differential Equations with Applications to Mathematical Physics*, vol. **192**. Academic Press, New York, 1993, pp. 287–305.

[133] Precise damping conditions for global asymptotic stability of nonlinear second-order systems, II (with P. Pucci). *J. Differential Equations*, **113** (1994), 505–534.

[134] Asymptotic stability for intermittently controlled nonlinear oscillators (with P. Pucci). *SIAM J. Math. Anal.*, **25** (1994), 815–835.

[135] On the derivation of Hamilton's equations (with P. Pucci). *Arch. Rational Mech. Anal.*, **125** (1994), 297–310.

[136] Ground states of a quasi-linear equation (with L.A. Peletier and H. Zou). *Differential Integral Equations*, **7** (1994), 1063–1082.

[137] Asymptotic estimates for a nonstandard second-order differential equation (with W.A. Harris and P. Pucci). In *Differential Equations, Dynamical Systems, and Control Science: A Festschrift in Honor of Lawrence Markus*, (K.D. Elworthy et al., eds.). Lecture Notes in Pure and Applied Mathematics. Marcel Dekker, New York, 1994, pp. 75–85.

[138] Classification of positive solutions of quasi-linear elliptic equations (with H. Zou). *Topological Meth. Nonlinear Anal.*, **3** (1994), 1–26.

[139] Asymptotic integration of second-order systems (with H. Zou). *Amer. J. Math.*, **116** (1994), 1241-1264.

[140] Bifurcation for semilinear equations at a singular limit (with F. Merle and L.A. Peletier). *Indiana Univ. Math. J.*, **43** (1994), 585–605.

[141] Non-existence of positive solutions of semilinear elliptic systems (with H. Zou). In *A Tribute to Ilya Bakelman*, (S.A. Fulling et al., eds.). Discourses in Mathematics and its Applications, vol. **3**. Department of Mathematics, Texas A&M University, 1994, pp. 56–69.

[142] Space, Time, and Energy. Lectio Doctoralis at the University of Ferrara. In this volume.

[143] Remarks on Lyapunov stability (with P. Pucci). *Differential Integral Equations*, **8** (1994), 1265–1278.

[144] A numerical study of the existence and non-existence of ground states and their bifurcation for the equations of Chipot and Weissler (with Y. Yan and H. Zou). Unpublished manuscript, 1994.

[145] Reflections on mathematics. Lectio Doctoralis at the University of Padova. In "*L'Anno Galileiano*," Vol. I. Ed. Lint, Trieste, 1995, pp. 120–124.

[146] Asymptotic stability for ordinary differential systems with time dependent restoring potentials (with P. Pucci). *Arch. Rational Mech. Anal.*, **132** (1995), 207–232.

[147] On the elementary thermodynamics of quasi-static systems and other remarks. In *Thermoelastic Problems and the Thermodynamics of Continua*. AMD-Vol. **198**, 53–62. Amer. Soc. Mech. Eng., Applied Mech. Div., 1995.

[148] Quasi-variational systems. In *First World Congress of Nonlinear Analysts* (V. Lakshmikantham, ed.). De Gruyter, Berlin, 1996, pp. 2055–2068.

[149] Asymptotic stability for non-autonomous damped wave systems (with P. Pucci). *Comm. Pure Appl. Math.*, **XLIX** (1996), 177–216.

[150] Existence and uniqueness of ground states of quasi-linear elliptic equations (with B. Franchi and E. Lanconelli). *Adv. in Math.*, **118** (1996), pp. 177–143.

[151] The Navier-Stokes equations and the laws of thermodynamics. *Meccanica*, **31** (1996), 546–563.

[152] Stability for abstract evolution equations (with P. Pucci). In *Partial Differential Equations and Applications*, (P. Marcellini et al., eds.). Marcel Dekker, New York, 1996, 279–288.

[153] Non-existence of positive solutions of Lane-Emden systems (with H. Zou). *Differential Integral Equations*, **9** (1996), 635–653.

[154] Global non-existence theorems for quasi-linear evolution equations with dissipation (with H. Levine). *Arch. Rational Mech. Anal.*, **137** (1997), 341–361.

[155] Asymptotic stability for nonlinear parabolic systems (with P. Pucci). In *Energy Mathods in Continuum Mechanics*. Kluwer Academic, Dordrecht, 1996, pp. 66–74.

[156] Existence of positive solutions of the Lane-Emden system (with H. Zou). *Atti Sem. Mat. Fis. Univ. Modena* (1997). To appear.

[157] Local asymptotic stability for dissipative wave systems (with P. Pucci). *Israel J. Math.* (1997). To appear.

[158] Stability and blow-up for dissipative evolution equations (with P. Pucci). In *Partial Differential Equations and Applications* (G. Caristi and E. Mitidieri, eds.). Lecture Notes in Pure and Applied Math., vol. **177**. Marcel Dekker, New York, 1977, pp. 279–288.

[159] Some remarks on global non-existence for non-autonomous abstract evolution equations (with H. Levine and P. Pucci). *Contemp. Math.*, **208** (1997), 253–263.

[160] Uniqueness of ground states for quasi-linear elliptic operators (with P. Pucci). To appear.

[161] The existence of positive entire solutions of elliptic Hamiltonian systems (with H. Zou). To appear.

[162] Global existence and nonexistence theorems for quasi-linear evolution equations of formally parabolic type (with H. Levine and S.R. Park), *J. Diff. Eqs.* To appear.

2
On Keplerian N-Body Type Problems

A. Ambrosetti and K. Tanaka

2.1. Introduction

The purpose of this chapter is to survey some recent results concerning the existence of periodic solutions of a class of systems of ordinary differential equations such as

$$\ddot{x} + V'(x) = 0, \tag{2.1.1}$$

where $x = (x_1, \ldots, x_N)$, $x_i \in \mathbf{R}^k$, and the potential V has the form

$$V(x) = \frac{1}{2} \sum_{1 \le i \ne j \le N} V_{ij}(x_i - x_j). \tag{2.1.2}$$

When V has the above form we will refer to (2.1.1) as an *N-body type system*. The name is motivated by the fact that if

$$V_{ij}(\xi) = -|\xi|^{-1}, \tag{2.1.3}$$

then (2.1.1) describes the motion of N particles in $\mathbf{R}^k$ with unitary masses, subjected to their Keplerian attraction.

More precisely, we will discuss here the existence of periodic solutions with prescribed energy h, namely, solutions to

$$\begin{cases} \ddot{x} + V'(x) & = & 0, \\ \frac{1}{2}|\dot{x}|^2 + V(x) & = & h, \end{cases} \tag{2.1.4}$$

with V of the form (2.1.2) and V_{ij} of the type

$$V_{ij}(\xi) \simeq -|\xi|^{-\alpha}, \qquad 0 < \alpha < 2. \tag{2.1.5}$$

This class of potentials will be called *Keplerian*.

Problem (2.1.4) is variational in nature and has been studied by means of critical point theory. The advantage of such an approach is that it does not use the specific features of the Kepler potential (2.1.3) but permits us to handle a broad class of potentials including those of the form (2.1.5).

While a certain number of results have been proved in the case $N = 2$, see [3, and references therein], much less is known when $N > 2$. In particular, we will review here the results of [2] and [4].

Let us recall that there are other results dealing with N-body type problems that will not be in this survey:

(1) the existence of solutions with prescribed period (see [5], [11], and [13]), that deal, roughly, with potentials V_{ij} of the form (2.1.5) with $\alpha \geq 2$; and [8], that is concerned with a class of Keplerian, *symmetric* potentials.

(2) the existence of multiple solutions for *restricted* problems: see [1] and [7].

(3) the estimate of the number of collisions: see [6] and [14].

Finally, we mention paper [12] where the existence of periodic solutions to (2.1.4) with $h > 0$ and $V_{ij}(\xi) \simeq -|\xi|^{-\alpha}$, $\alpha > 2$, is proved.

2.2. The Functional Setting

In the sequel, we will deal with potentials of the form (2.1.2) with $V_{ij} \in C^1(\mathbf{R}^k \setminus \{0\}, \mathbf{R})$ (of class C^2 in Theorem 2.3) and satisfying (2.1.5).

(a) The Variational Principle

Let us consider the Sobolev space $H = H^{1,2}(S^1, \mathbf{R}^k)$. For $u_i \in H$ we set $[u_i] = \int_0^1 u_i(s)\,ds$ and consider the Hilbert space $E = \{u \in H^N : \sum [u_i] = 0\}$. In the sequel, we will use the notation $\|\cdot\|_2$ to denote the L^2-norm; moreover we will write $u^n \rightharpoonup u$ to indicate that the sequence u^n converges to u weakly in E and uniformly on $[0, 1]$. Let

$$\Lambda = \{u \in E : u_i(t) \neq u_j(t),\ \forall t \in S^1\}$$

and, for $u \in \Lambda$, define

$$I(u) = \frac{1}{2}\|\dot{u}\|_2^2 \cdot \int_0^1 (h - V(u))\,dt.$$

The following is essentially the Maupertuis Principle (MP).

(MP) Let $u \in \Lambda$ be such that $I(u) > 0$ and $I'(u) = 0$, and set

$$\omega^2 = \frac{4I(u)}{\|\dot{u}\|_2^4}. \tag{2.2.1}$$

Then $x(t) = u(\omega t)$ is a classical solution of (2.1.4).

Let us remark explicitly that such an x is such that $x_i(t) \neq x_j(t)$ for all $t \in \mathbf{R}$, and will be referred to as a *noncollision solution.*

(b) Strong Forces and Keplerian Potentials

It is clear that in order to find critical points of I on Λ we have, first of all, to control the behavior of I on the boundary $\partial\Lambda$, consisting of those $u \in E$ such that $u_i(t) = u_j(t)$ for some $t \in \mathbf{R}$. In this connection, we have the following:

Lemma 2.1. *Suppose that* $\exists\, c, r > 0$ *such that*

$$V_{ij}(\xi) \leq -\frac{c}{|\xi|^2}, \qquad \forall 0 < |\xi| < r, \qquad \forall i \neq j. \tag{2.2.2}$$

Then, for any sequence $u^n \in \Lambda$ *such that* $u^n \rightharpoonup u \in \partial\Lambda$, *there results*

$$\int_0^1 V(u^n)\, dt \to -\infty. \tag{2.2.3}$$

Condition (2.2.2) has been introduced by Gordon [9] and is usually referred to as the *Strong Force Condition.*

In contrast with the preceding lemma, when (2.1.5) holds, it may happen that $\int_0^1 V(u^n)\, dt$ remains bounded as $u^n \rightharpoonup u \in \partial\Lambda$. Actually, this is the case when $V_{ij}(\xi) = -|\xi|^{-1}$ and u is a collision solution of (2.1.4). In order to overcome this difficulty, we will follow a procedure systematically used when $N = 2$. We consider a perturbed potential of the form

$$V_\epsilon(u) = V(u) - \epsilon \cdot \sum |u_i - u_j|^{-2} \tag{2.2.4}$$

and the corresponding perturbed Maupertuis-like functional

$$I_\epsilon(u) = \frac{1}{2}\|\dot{u}\|_2^2 \cdot \int_0^1 (h - V_\epsilon(u))\, dt. \tag{2.2.5}$$

According to (2.2.3) we have that

$$\int_0^1 \sum_{i \neq j} |u_i^n - u_j^n|^{-2}\, dt \to \infty, \tag{2.2.6}$$

whenever $u^n \rightharpoonup u \in \partial\Lambda$ and this will be useful in controlling the behavior of I_ϵ near $\partial\Lambda$.

As in the (MP), let $u^\epsilon \in \Lambda$ be any critical point of I_ϵ such that $c_\epsilon = I_\epsilon(u^\epsilon) > 0$ and set

$$\omega_\epsilon^2 = \frac{4c_\epsilon}{|\dot{u}^\epsilon|_2^4}.$$

Then $x^\epsilon(t) = u^\epsilon(\omega_\epsilon t)$ is a noncollision solution of the perturbed problem

$$\begin{cases} \ddot{x} + V_\epsilon'(x) & = & 0, \\ \frac{1}{2}|\dot{x}|^2 + V_\epsilon(x) & = & h. \end{cases} \tag{2.2.7}$$

Solutions of (2.1.4) will be found by a limiting procedure. If for all $0 < \epsilon \leq \epsilon_0$ $x^\epsilon \in \Lambda$ is a solution of (2.2.7),

$$x^\epsilon \rightharpoonup x \qquad (\epsilon \to 0),$$

and

$$\exists \tau \quad \text{such that} \quad x_i(\tau) \neq x_j(\tau), \qquad \forall i \neq j, \tag{2.2.8}$$

we say that x is a *weak solution* of (2.1.4), see [3, Def. 10.1]. According to Theorem 10.7 of [3], any weak solution has the properties that:

(1) $\Gamma = \Gamma(x) = \{t \,:\, x_i(t) = x_j(t), i \neq j\}$ has measure zero; and

(2) x is a classical solution of (4) on $\mathbf{R} \setminus \Gamma$.

In the sequel we will take *negative* values of the energy. This is motivated by the fact that in the model case $V_{ij}(\xi) = -|\xi|^{-\alpha}$ and $0 < \alpha < 2$, problem (2.1.4) has a solution iff $h < 0$. It is worth pointing out that the "natural" value of the energy is *positive* when $V_{ij}(\xi) \simeq -|\xi|^{-\alpha}$ with $\alpha \geq 2$. Note also that the sign of h effects the behavior of I: if $h > 0$, the functional I is bounded below, while if $h < 0$, then I (as well as I_ϵ) is unbounded.

2.3. Symmetric Potentials

The first result we will discuss concerns the case in which

$$V_{ij}(\xi) = V_{ji}(\xi), \qquad \forall \xi \in \mathbf{R}^k \setminus \{0\}. \tag{2.3.1}$$

If (2.3.1) holds, then it follows that $V(x) = V(-x)$. To take advantage of this symmetry, we consider the subspace

$$E_0 = \{u \in H^N \,:\, u(t + 1/2) = -u(t)\}$$

and

$$\Lambda_0 = \{u \in E_0 \,:\, u_i(t) \neq u_j(t),\ \forall t \in S^1\}.$$

It turns out that any critical point $u \in \Lambda_0$ of I_ϵ constrained on E_0 is a critical point of I_ϵ on E and hence a solution of (2.2.7). This has first been used in [8] to prove the existence of solutions to (2.1.1) with prescribed period. Let us remark that no constant function but $u \equiv 0$ belongs to E_0. Hence for $u \in E_0$, $\|\dot{u}\|_2$ is a norm equivalent to the usual H^1 norm.

Let us suppose that:

(V1) $\exists \alpha \in [1, 2)$ such that $V'_{ij}(\xi) \cdot \xi \geq -\alpha V_{ij}(\xi) > 0, \quad \forall \xi \neq 0$;

(V2) $\exists r_1 > 0, \ \beta \in (0, 2)$ such that $V'_{ij}(\xi) \cdot \xi \leq -\beta V_{ij}(\xi), \quad \forall 0 < \xi < r_1$; and

(V3) $V_{ij}(\xi) \to 0$ as $|\xi| \to \infty$.

It is clear that $V_{ij}(\xi) = -|\xi|^{-\alpha}$, $1 \leq \alpha < 2$, satisfies the above assumptions, so that the Kepler interaction potential (2.1.3) fits in this frame.

Theorem 2.2. ([2]). *Let V have the form* (2.1.2) *and suppose that V satisfy* (2.3.2) *and* (V1,2,3). *Then for all $h < 0$ problem* (2.1.4) *has a weak periodic solution.*

According to the discussion in Section 2.2(b), and the remarks at the beginning of this section, we consider the perturbed functional I_ϵ defined in (2.2.5) and look for its critical points on Λ_0. The main steps of the proof are:

(1) existence of critical points of I_ϵ by means of the Mountain -Pass Theorem; and

(2) limiting process as $\epsilon \to 0$.

As for the former, we first show that the functional I_ϵ has the Mountain Pass geometry, in the sense that:

(i) $\exists \rho > 0, \ a > 0$, such that $I_\epsilon(u) \geq a$ for all $u \in \Lambda_0, \ \|\dot{u}\|_2 = \rho$;

(ii) $\exists u_0 \in \Lambda_0, \ \|\dot{u}_0\|_2 < \rho$, such that $I_\epsilon(u_0) < a$; and

(iii) $\exists u^* \in \Lambda_0, \ \|\dot{u}^*\|_2 > \rho$, such that $I_\epsilon(u^*) \leq 0$.

In particular, this latter statement depends on the fact that $h < 0$ and (V3) holds. Moreover we also prove that $(\mathrm{PS})_c$ holds for all $c > 0$, namely, that any sequence u^n such that

$$\begin{cases} (i) & I_\epsilon(u^n) \to c, \\ (ii) & I'_\epsilon(u^n) \to 0, \end{cases} \tag{2.3.2}$$

converges, up to a subsequence, to some $u \in \Lambda_0$. To prove this fact, we first use (2.3.2) to infer that $u^n \rightharpoonup u$. Let us remark that although, in general, there could be sequences verifying (i) and (ii) which converge to $u \equiv 0$, this is not the case when (V1) holds with $\alpha \geq 1$. Actually, this implies that

$$\limsup I_\epsilon(u^n) \leq 0,$$

whenever $u^n \rightharpoonup 0$. Since $u \not\equiv 0$ then, using (2.2.6), we infer that $u \in \Lambda_0$ and $(\mathrm{PS})_c$ follows.

As for Step 2, we first derive some uniform estimates, proving there exist constants c, Ω', $\Omega'' > 0$ such that

$$\|\dot{u}^\epsilon\|_2 \le c,$$
$$\Omega' \le \omega_\epsilon \le \Omega'',$$

and hence $u^\epsilon \rightharpoonup u$, $\omega_\epsilon \to \omega \neq 0$. Setting $x(t) = u(\omega t)$ we also show that (2.2.8) holds and this suffices to say that x is a weak solution of (2.1.4).

For more details and other results, we refer to [2].

2.4. The General Case

In this section we deal with potentials which are not symmetric. In this case, we meet the additional difficulty that the (PS) condition breaks down. For example, let $N = 3$, $V_{ij}(\xi) = -|\xi|^{-\alpha}$, and let $y = (y_1, y_2)$ be a circular periodic solution of the 2-body problem with energy $h_{1,2} < 0$ and such that $[y_1]+[y_2] = 0$. Consider a sequence $\xi^n \in \mathbf{R}$ such that $|\xi^n| \to \infty$ and let $x^n = (y_1+\xi^n, y_2+\xi^n, -2\xi^n) \in \Lambda$. Such an x^n is a nonconverging sequence verifying (2.3.2(i), (ii)), with

$$c = \frac{1}{2}\left(\|\dot{y}_1\|_2^2 + \|\dot{y}_2\|_2^2\right) \cdot \int_0^1 (h_{1,2} + |y_1 - y_2|^{-\alpha})\, dt.$$

Moreover it is easy to see that this strong lack of compactness survives even if we deal with I_ϵ. To overcome this problem, Morse theoretical arguments have been used by Bahri and Rabinowitz in [5] (in the frame of the existence of solutions of (2.1.1) with fixed period). A different approach has been introduced in [11] (still for the fixed period problem). Such a method has been subsequently employed in [12] to prove the existence of solutions of (2.1.4) with $h > 0$ and, roughly, $V_{ij} \simeq -|\xi|^\alpha$ with $\alpha > 2$.

Similar arguments have been carried over in a recent paper [4] dealing with Keplerian-like potentials, and we are going to survey below the results therein.

We suppose that $V_{ij} \in C^2(\mathbf{R}^k \setminus \{0\}, \mathbf{R})$ satisfies (V2) and

(V1') $\exists \alpha \in (0, 2) \; : \; V'_{ij}(\xi) \cdot \xi \ge -\alpha V_{ij}(\xi) > 0, \; \forall \xi \neq 0$;

(V4) $\lim_{|\xi|\to\infty}(V_{ij}(\xi) + \frac{1}{2} V'_{ij}(\xi) \cdot \xi) = 0$;

(V5) $\exists \theta \in [0, \pi/2), \; r_2 > 0 \; : \; \mathrm{ang}(V'_{ij}(\xi), \xi) \le \theta, \; \forall |\xi| > r_2$; and

(V6) $3V'_{ij}(\xi) \cdot \xi + V''_{ij}(\xi)\xi \cdot \xi \neq 0, \; \forall \xi \neq 0$.

In (V5), $\mathrm{ang}(\eta, \xi)$ denotes the angle between the vectors η and ξ. For future reference we note that (V1', 2) imply that $\exists c, r > 0$ such that

$$V_{ij}(\xi) \le -c|\xi|^{-\alpha}, \qquad \forall 0 < |\xi| < r; \tag{2.4.1}$$

$$\lim_{|\xi|\to 0}(V_{ij}(\xi)+\frac{1}{2}V_{ij}'(\xi)\cdot\xi)=-\infty. \tag{2.4.2}$$

Our main result is:

Theorem 2.3. ([4]). *Let V have the form* (2.1.2) *and suppose that* $V_{ij}\in C^2(\mathbf{R}^k\setminus\{0\},\mathbf{R})$ *satisfy* (V1',2,4,5,6). *Then for all* $h<0$ (2.1.4) *has a weak periodic solution.*

The proof of Theorem 2.3 is carried out in several steps that we will outline below.

Step 1. (*An Auxiliary Functional*). Using (2.4.2), (V4), and (V6), it follows that for all $u\in\Lambda$, $\dot u\not\equiv 0$, the equation

$$\frac{d}{d\rho}I_\epsilon(\rho u)=0$$

has a unique solution $\rho=\rho(u)>0$, which is independent of ϵ, satisfying

$$h=\int_0^1[V(\rho u)+\frac{1}{2}V'(\rho u)\rho u]\,dt. \tag{2.4.3}$$

Let

$$J_\epsilon(u)=I_\epsilon(\rho(u)u)$$

and consider the manifold $M=\{u\in\Lambda\ :\ \|\dot u\|_2\doteq 1\}$. For $u\in M$ we have

$$J_\epsilon(u)=\frac{1}{4}\rho^2\int_0^1 V'(\rho u)\rho u\,dt+\frac{1}{2}\epsilon\int_0^1\sum|u_i-u_j|^{-2}\,dt. \tag{2.4.4}$$

As a consequence of (2.4.4) and (V1'), it follows that $J_\epsilon>0$ on M. Moreover any $u^\epsilon\in M$ such that $\nabla_M J_\epsilon(u^\epsilon)=0$ (hereafter $\nabla_M J_\epsilon$ denotes the constrained gradient of J_ϵ on M) satisfies $I_\epsilon'(\rho_\epsilon u^\epsilon)=0$ and hence gives rise, through (MP), to a solution of (2.2.7).

Step 2. (*The (PS) Condition*). As anticipated before (I_ϵ and hence) J_ϵ does not satisfy the (PS) condition. To overcome such a problem, we introduce, following [11] and [12], the auxiliary functional

$$g(u)=\sum_{i\neq j}|[u_i-u_j]|^2$$

and consider the manifold with boundary

$$M^b=\{u\in M\ :\ g(u)\le b\}.$$

In order to show that J_ϵ satisfies (PS) on M^b we first prove:

Lemma 2.4.

(i) $\forall c > 0, \exists k_1 > 0$ such that $J_\epsilon(u) \leq c \Longrightarrow \rho(u) \leq k_1, \ \forall u \in M, \ \forall \epsilon \in (0, 1]$;

(ii) $\forall c > 0, \ \exists k_2 > 0$ such that $J_\epsilon(u) \geq c \Longrightarrow \rho(u) \geq k_2, \ \forall u \in M, \ \forall \epsilon \in (0, 1]$; and

(iii) $u^n \rightharpoonup u \in \Lambda \Longrightarrow \rho(u^n) \to \rho(u)$.

From Lemma 2.4 it follows that:

Lemma 2.5.

(i) Suppose that $u^n \in M^b$ is such that $J_\epsilon(u^n) \to c > 0$ and $\nabla_M J_\epsilon(u^n) \to 0$. Then, up to a subsequence, $u^n \to u \in M$, strongly.

(ii) Suppose that $u^n \in M$ is such that $J_\epsilon(u^n) \leq c$, $g(u^n) = b$, and $\nabla_M J_\epsilon(u^n) - \mu_n \nabla_M g(u^n) \to 0$ $(\mu_n > 0)$. Then, up to a subsequence, $u^n \to u \in M$, strongly.

The proof relies, roughly, on the fact that:

(i) since M^b is bounded in E, then, up to a subsequence, $u^n \rightharpoonup u$;

(ii) we show that such an u is collision-free, namely, $u \in \Lambda$ because, otherwise, (2.2.6) would imply

$$\lim J_\epsilon(u^n) \geq \frac{1}{2}\epsilon \cdot \lim \int_0^1 \sum |u_i^n - u_j^n|^{-2}\, dt = +\infty;$$

and

(iii) using Lemma 2.4 we deduce that $u^n \to u$.

Step 3. (*Approximate Solutions*). In order to find critical points of J_ϵ on M^b we will use, as in [12], the theory of critical points on a manifold with the boundary developed in [10]. Referring to [10] for more details, we recall that in such a paper critical points of a functional F on a manifold $\mathcal{M}$ with boundary $\partial\mathcal{M}$ are searched. Roughly, it is first shown that:

(M(i)) the steepest descent method works provided F has no critical points constrained on $\partial\mathcal{M}$ which are "outward pointing".

If this is the case, a critical point of F in $\mathcal{M}$ exists whenever

(M(ii)) $\mathcal{M}$ cannot be deformed into $\partial\mathcal{M}$.

Let us begin with point (M(ii)) and consider, again following [12], the class $\mathcal{A}$ of *admissible sets*: we say that $A \subset \Lambda$ is admissible if for any homotopy $\eta \in C([0,1] \times E, E)$ with$\eta(0,\cdot) = \mathrm{id.}$, $\eta(1, A) \subset \mathbf{R}^{kN}$, A contains at least one point u with the property that $\forall\, i \neq j, \exists\, i_1, \ldots, i_m \in \{1, \ldots, N\}$ such that

(1) $i_1 = i,\ i_m = j,\ i_k \neq i_{k+1},\ \forall k = 1, \ldots, m-1$; and

(2) $\eta([0,1] \times \{u\}) \cap \Gamma(i_k, i_{k+1}) \neq \emptyset,\ \forall k$,

where $\Gamma(i_k, i_{k+1}) = \{u \in E\ :\ u_{i_k}(t) = u_{i_{k+1}}(t)$ for some $t\}$.
The following properties hold

$$A \subset B, \qquad A \in \mathcal{A} \quad \Rightarrow \quad B \in \mathcal{A}; \tag{2.4.5}$$

$$A \in \mathcal{A} \text{ and } B \text{ is a deformation of } A \Rightarrow B \in \mathcal{A}. \tag{2.4.6}$$

We set $\mathcal{A}_M = \{A \subset M\ :\ A \in \mathcal{A}\}$. It is known that there exists $A \in \mathcal{A}$ which does not contain any identically constant function. Plainly the radial projections on M of such an A belongs to $\mathcal{A}_M$ which is therefore not empty.

Using admissible sets we define

$$c_\epsilon^* = \inf_{A \in \mathcal{A}_M} \sup_{u \in A} J_\epsilon(u).$$

Plainly, we have that $c_\epsilon^* \leq c_1^*$. The definition of $\mathcal{A}_M$, properties (2.4.5–2.4.6) and Lemma 2.4 yield:

Lemma 2.6.

(i) *There exists* $b' > 0$ *such that* $\{u \in M\ :\ g(u) \geq b'\} \notin \mathcal{A}_M$; *and*

(ii) *for all* $b' > 0$ *there exists* $\gamma = \gamma(b') > 0$, *independent of* ϵ, *such that* $\{u \in M\ :\ J_\epsilon(u) \leq \gamma\} \subset \{u \in M\ :\ g(u) \geq b'\}$.

Coming back to point (M(i)) above, we show, also using (V5), that the following lemma holds:

Lemma 2.7. *For all* $0 < c \leq C$ *there exists* $b = b(c, C) > 0$, *independent of* ϵ, *such that*

$$\nabla_M J_\epsilon(u) = \lambda \nabla_M g(u), \qquad \lambda > 0,$$

has no solutions on $\{u \in M\ :\ c \leq J_\epsilon(u) \leq C\} \cap \{u \in M\ :\ g(u) \geq b\}$.

Fixed any $C^* > c_1^*$, we take γ as in Lemma 2.6 and $b'' = b(\gamma, C^*)$ according to Lemma 2.7. We set

$$G(a, b) = \{u \in M\ :\ J_\epsilon(u) \leq a\} \cup \{u \in M\ :\ g(u) \geq b\}.$$

Letting $b^* = \max\{b', b''\}$ and using Lemma 2.6(ii) we infer that

$$G(\gamma, b^*) \subset \{u \in M \,:\, g(u) \geq b'\}.$$

Then Lemma 2.6(i) and property (2.4.5) imply that

$$G(\gamma, b^*) \notin \mathcal{A}_M. \tag{2.4.7}$$

Moreover, the definition of c_ϵ^* yields

$$G(C^*, b^*) \in \mathcal{A}_M. \tag{2.4.8}$$

Finally, we set

$$c_\epsilon = \inf\{c \in \mathbf{R} \,:\, G(c, b^*) \in \mathcal{A}_M\}.$$

Statements (2.4.7)–(2.4.8), Lemmas 2.5–2.7, jointly with the preceding discussion, leads us to show that c_ϵ is a critical value of J_ϵ. Precisely we have:

Lemma 2.8. *There eists $u^\epsilon \in M^{b^*}$ such that $\nabla_M J_\epsilon(u^\epsilon) = 0$ with $c_\epsilon = J_\epsilon(u^\epsilon) \in [\gamma, C^*]$.*

According to the (MP) principle, it follows that, letting $\rho_\epsilon = \rho(u^\epsilon)$ and $v^\epsilon = \rho_\epsilon u^\epsilon$,

$$x^\epsilon(t) = v^\epsilon(\omega_\epsilon t), \qquad \omega_\epsilon = 2\sqrt{I_\epsilon(v^\epsilon)} \cdot \rho_\epsilon^{-2}, \tag{2.4.9}$$

is a solution of (2.2.7).

Step 4 (*Limiting Process*). Since $u^\epsilon \in M^{b^*}$ then, up to a subsequence, $u^\epsilon \rightharpoonup u \in E$. Moreover, using Lemma 2.4 and the fact that $\gamma \leq c_\epsilon \leq C^*$, we show:

Lemma 2.9.

(i) $\exists\, a_0,\ a_1 > 0$ *such that* $a_0 \leq \rho_\epsilon \leq a_1$, $\forall \epsilon > 0$ and hence $\rho_\epsilon \to \rho \neq 0$; and

(ii) $\omega_\epsilon \to \omega \neq 0$.

Letting $x(t) := \rho u(\omega t)$, it follows that x^ϵ, defined in (2.4.9), converges weakly to x. Moreover we show there exists $\tau \in [0, 2\pi/\omega]$ such that $x_i(\tau) \neq x_j(\tau)$ for all $i \neq j$ and this suffices to say that x is a weak solution of (2.1.4), proving Theorem 2.3.

References

[1] Ambrosetti, A. and Bessi, U. *Multiple closed orbits for perturbed Keplerian problems. J. Differential Equations*, **96** (1992), 283–294.

[2] Ambrosetti, A. and Coti Zelati, V. *Closed orbits with fixed energy for a class of N-body problems. Ann. Inst. H. Poincaré Anal. NonLinéaire*, **9** (1992), 187–200 and 337–338.

[3] Ambrosetti, A. and Coti Zelati, V. *Periodic Solutions of Singular Lagrangian Systems.* Birkhäuser, Boston, 1993.

[4] Ambrosetti, A. and Tanaka, K., and Vitillaro, E. Periodic solutions with prescribed energy for some Keplerian N-body problems. Preprint. Scuola Normale Superiore Pisa, 1993.

[5] Bahri, A. and Rabinowitz, P.H., *Solutions of the three-body problem via critical point at infinity. Ann. Inst. H. Poincaré Anal. NonLinéaire.* **8** (1991), 561–649.

[6] Bessi, U. and Coti Zelati, V. *Symmetries and noncolusion closed orbits for noncollision planar N-body type problems. Nonlinear Anal. T.M.A., 16 (1991), 587–598.*

[7] Coti Zelati, V. A class of periodic solutions of the N-body problem. Celestial Mech., **46** (1989), 177–186.

[8] Coti Zelati, V. *Periodic solutions for the N-body type problems. Ann. Inst. H. Poincaré Anal. NonLinéaire*, **7** (1990), 477–492.

[9] Gordon. *Conservative dynamical systems involving strong forces. Trans. Amer. Math. Soc.*, **204** (1975), 113–135.

[10] Majer, P. *Two variational methods on manifolds with boundary.* Topology, 34-1 (1995), 1–12.

[11] Majer, P. and Terracini S. *Periodic solutions to some problems of N-body type. Arch. Rational Mech. Anal.* **124** (1993), 683–697.

[12] Majer, P. and Terracini, S. *Periodic solutions to some N-body problems: the fixed energy case. Duke Math. J.* **69** (1993), 683–697.

[13] Riahi, H. Periodic orbits of N-body type problems. Ph.D. Thesis, Rutgers University, New Brunswick, 1992.

[14] Serra, E. and Terracini, S. *Arch. Rational Mech. Anal.*, **120** (1992), 305–325.

3
Invariance and Balance in Continuum Mechanics

G. Capriz and G. Mazzini

3.1. Introduction

The search for a convenient set of axioms for classical continuum mechanics has interest not only "per se" (Hilbert's sixth problem), but also because the choice of a particular set may suggest, or hinder, generalizations appropriate to circumstances more complex that those envisaged classically (e.g., to mixtures, continua with microstructure, etc.).

The axioms suggested by Noll [1] are declarations of properties of interactions between subbodies of a given body; they lead, through an elegant and rigorous analysis, to corollaries regarding fields of stress and their modes of evolution. The successive path to perspectives ampler than classical is, however, ardous or hazardous: we need only quote the delicacy of the analysis, by Noll and Virga, of edge interactions [2] and the fragility of the proposal, by Capriz and Virga, to use embeddings in higher-dimensional linear spaces [3].

Many authors begin with the concept of virtual work and assume certain properties of formal invariance to deduce evolution equations. With such a point of departure, the path to wider vistas appears almost too easy, owing perhaps to the fact that the notion of stress is presumed rather than derived; here the wealth of possible citations is almost bewildering. Germain has given concise reports and suggested many extensions (see, e.g., [4]); for an intrinsic formulation, see [5].

After listing arguments in favor of this second approach, Truesdell and Toupin in [6, p. 594ff] rebut them and complain, in particular, that "no variational principle has ever been shown to yield Cauchy's fundamental theorem in its basic sense as asserting that the existence of a stress vector implies the existence of a stress tensor."

We propose, below, a course which, though it originates from the principle of virtual power, seems to avoid the main criticism of Truesdell and Toupin and, still, to open the way toward interesting generalizations. In this vein, we quote explicitly continua with couple stresses and continua with microstructure; to deal with the latter, advantage is taken of studies by Segev [7].

On the whole, our present contribution is one of intensity and detail, but we hope to show, in later papers, that here is the seed for new nontrivial developments.

Needless to say, we are encouraged in our endeavors by lectures (on the tight relationship between conditions of balance and of invariance) given by Professor Serrin (see, e.g., [8]); to him we are honored to dedicate this chapter.

Note that, although we could have used a totally intrinsic material setting (such as Noll's [9] or Rougée [5]), we adopt below the "naive" setting. Also, to be brief, and at the cost of leaving open the question of the most general form of the boundary conditions, we consider below tacitly only subbodies of a body, which is itself totally internal to a larger body of the same material type.

3.2. The Principle of Virtual Power

Let $\mathcal{B}_*$ be a region of a three-dimensional Euclidean space $\mathcal{E}$, with translation space $\mathcal{V}$; for later convenience assume $\mathcal{B}_*$ to be regular (i.e., to be a three-dimensional, compact, connected C^2–manifold embedded in $\mathcal{E}$). Imagine $\mathcal{B}_*$ as a convenient placement of a physical body, so that any other placement $\mathcal{B}$ of interest can be obtained through a continuous displacement field $u \in C(\mathcal{B}_*; \mathcal{V})$: $\mathcal{B} = \{x | x(y) = y + u(y);\ y \in \mathcal{B}_*\}$. Inappropriately we will speak of the "body $\mathcal{B}$" or of the "subbody $\mathfrak{b}$ of the body $\mathcal{B}$." This ambiguity does not harm the analysis and, we hope, will not distress unduly the reader to whom we leave a modest task of interpretation.

Let

$$\mathcal{P} := \{x \in C^0(\mathcal{B}_*; \mathcal{E}) | x(\mathcal{B}_*) \text{ has the same regularity as } \mathcal{B}_*\}$$

be a suitable space of all placements of the body. Let $\mathcal{B} = x(\mathcal{B}_*)$ and attach to any $x \in \mathcal{P}$ a field v of virtual velocities and define

$$\mathcal{T}_x\mathcal{P} := \{(x, v) | v \in C^0(\mathcal{B}, \mathcal{V})\},$$

so that, for any fixed $\hat{x} \in \mathcal{P}$, $\mathcal{T}_{\hat{x}}\mathcal{P}$ can also be identified with $C^0(\mathcal{B}_*; \mathcal{V})$.

By a known integral representation theorem (see, e.g., [10], Th. 19.9) there is a natural isomorphism between the dual space $\mathcal{T}_x^*\mathcal{P}$ of $\mathcal{T}_x\mathcal{P}$ and the space of vector valued measures on $\mathcal{B}_*$ (or, equivalently, $\mathcal{B}$); we interpret any element of $\mathcal{T}_x^*\mathcal{P}$ as a 'working', i.e. as a virtual power $\pi_{\mathcal{B}}$. In conclusion, we may say that: For any field of virtual velocities v attached to a placement x, there correspond workings $\pi_{\mathcal{B}}(v)$, $\pi_{\mathcal{B}} \in \mathcal{T}_x^*\mathcal{P}$; for any one of the workings there exist a vector–valued measure $m_{\mathcal{B}}$ such that

$$\pi_{\mathcal{B}}(v) = \int_{\mathcal{B}} v(x) \cdot dm_{\mathcal{B}}(x). \tag{3.2.1}$$

Formally the result can be applied to any appropriate subbody $\mathfrak{b}$ of the body $\mathcal{B}$, simply by writing $\mathfrak{b}$ instead of $\mathcal{B}$ in (3.2.1), but notice that the measure will depend also on $\mathfrak{b}$. Actually it is the form of that dependence, that becomes the critical issue, when modelling the behaviour of classes of bodies. Some essential properties need be satisfied when a special form is proved directly. The fundamental one among those properties is the property of additivity: let $\mathfrak{b}_i (i = 1, 2, \ldots, r)$ be any finite set of disjoint subbodies of $\mathcal{B}$, the union (or rather the join) of which coincides with $\mathcal{B}$ itself; then

$$\pi_{\mathcal{B}}(v) = \sum_{i=1}^{r} \pi_{\mathfrak{b}_i}(v). \tag{3.2.2}$$

In classical continuum mechanics, we assume that there exist fields $b_{\mathfrak{b}}(x) \in C^0(\mathfrak{b}; \mathcal{V})$ and $t_{\mathfrak{b}}(x) \in C^0(\partial\mathfrak{b}; \mathcal{V})$ such that

$$\pi_{\mathfrak{b}}^{\mathrm{cl}}(v) = \int_{\mathfrak{b}} v(x) \cdot b_{\mathfrak{b}}(x) d\,(\mathrm{vol}) + \int_{\partial\mathfrak{b}} v(x) \cdot t_{\mathfrak{b}}(x) d\,(\mathrm{area}), \tag{3.2.3}$$

i.e., the measure is split into the sum of a measure which is absolutely continuous with respect to the volume of $\mathfrak{b}$ and one absolutely continuous with respect to the area of $\partial\mathfrak{b}$. Then sufficient conditions that assure compatibility of (3.2.3) with (3.2.2) are the following:

(i) the field $b_{\mathfrak{b}}(x)$ is independent of $\mathfrak{b}$;

(ii) $t_{\mathfrak{b}}(x)$ depends on $\mathfrak{b}$ only through the outer unit normal $n_{\mathfrak{b}}(x)$ to $\partial\mathfrak{b}$ at x and is an odd function of $n_{\mathfrak{b}}$.

It should be noted that if virtual velocities were choosen in $C^1(\mathcal{B}_*; \mathcal{V})$, a more general representation theorem could be applied to get *stresses* which do not necessarily derive from *tractions* (when they have no trace, i.e. no local action exists): that is the case, for instance, of Maxwell's tensor.

The choice (3.2.3) may appear, in the present context, very special, although experimental evidence shows it to be almost exhaustive. That it is not completely so, is indicated already by Toupin's line distribution of forces and Vianello's suggestion of edge interactions between bodies [11], interactions precisely defined and deeply explored by Noll and Virga [2]. They have specified further the notion of regular region (and of its regular partitions, comprising sides, edges and vertices of its border), of simple and of essential contact between two subbodies of a given body and finally come to a proposal which, in our present context, can be expressed as follows: for certain bodies the measure $m_{\mathfrak{b}}$ is such that $\pi_{\mathfrak{b}}$ is the sum of three terms: the two already accounted for in (3.2.3) plus one which involves a force per unit length of edges $e^{(i)}$, if present,

$$\pi_{\mathfrak{b}}^{\mathrm{cs}}(v) = \pi_{\mathfrak{b}}^{\mathrm{cl}}(v) + \int_{\Sigma e^{(i)}} v(x) \cdot f_{\mathfrak{b}}(x) d\,(\mathrm{length}). \tag{3.2.4}$$

The careful analysis of Noll and Virga allows them to come to a specification of $b_{\mathfrak{b}}$, $t_{\mathfrak{b}}$ and $f_{\mathfrak{b}}$ that satisfies (3.2.2) under otherwise general circumstances. The matter is very delicate because of the possible interplay among the three integrals and because $t_{\mathfrak{b}}$ and $f_{\mathfrak{b}}$ depend now more deeply on the geometric properties of $\partial\mathfrak{b}$.

Working (in the example above: volume, surface and line working) may refer to a part or to all external actions on the body. There is a fundamental property of the global external working on which the whole of our analysis rests. The property is declared in an axiom, often refered to as Noll's axiom, which we proceed to adapt to our present circumstances. A necessary premiss would be the exact specification of the set of placements we consider; the matter is "likely to be extremely difficult and to require concepts from geometric measure theory that have not yet been invented" ([2], Introduction). Besides, we need to choose a setting: "all the massy bodies in the universe" or "those bodies which have occupied a portion of the event world within a limited period of time" ([12], I. 13). When the second choice is made (as when we include only all subbodies of a given body), an axiom of inertia needs be added, perhaps in a struthious manner under the apparently easier guise of a hypothesis of quasi balance.

We leave all these, though fundamental, matters aside and accept a version of Noll's axiom in the hope that an adeguate, usefully ample, reduced universe of bodies and an appropriate delimitation of external actions can be found where it applies.

Noll's Axiom. *For any choice of* $\mathfrak{b}$ *in the universe of subbodies of a given body* $\mathcal{B}$*, the global working* $\pi_{\mathfrak{b}}$ *of all actions on* $\mathfrak{b}$*, external to* $\mathfrak{b}$*, is frame indifferent.*

Because for any rigid motion of $\mathfrak{b}$ there is a frame $\mathcal{F}$ in which $\mathfrak{b}$ is at rest, Noll's axiom can be read as follows: in $\mathcal{F}$ the global external working on any $\mathfrak{b}$ is null.

3.3. Balance Equations

If we adopt (3.2.1) to the case of the global power of all external actions on any subbody $\mathfrak{b}$ and invoke Noll's axiom, we come to the condition

$$c \cdot \int_{\mathfrak{b}} dm_{\mathfrak{b},\text{ext}} + w \cdot \int_{\mathfrak{b}} x \times dm_{\mathfrak{b},\text{ext}} = 0, \quad \forall c, w \in \mathcal{V}, \quad \forall \mathfrak{b} \subset \mathcal{B},$$

(where c and w are translational and angular velocity in a rigid motion) and finally to the balance equations of momentum and moment of momentum

$$\int_{\mathfrak{b}} dm_{\mathfrak{b},\text{ext}} = 0 \qquad \int_{\mathfrak{b}} x \times dm_{\mathfrak{b},\text{ext}} = 0 \qquad \forall \mathfrak{b} \subset \mathcal{B}. \tag{3.3.1}$$

We cannot go further without some assumptions on $m_{\mathfrak{b},\text{ext}}$, even though they be relatively mild. For instance, if (3.2.3) applies, b is assumed to be independent of $\mathfrak{b}$ and $t_{\mathfrak{b}}(x)$ to depend on $\mathfrak{b}$ only through the direction $n_{\mathfrak{b}}(x)$ of the normal to $\partial\mathfrak{b}$ at x and to satisfy modest qualitative requirements, the existence of the stress tensor T follows. The integrals in (3.3.5) can be reduced to integrals over $\mathfrak{b}$ with a volume-measure; the arbitrariness of $\mathfrak{b}$ can be exploited and the usual local balance conditions derived

$$\operatorname{div} T + b = 0, \qquad T \in \mathrm{Sym}. \tag{3.3.2}$$

The case when (3.2.4) applies is much more delicate; we refer the reader to [2] for an accurate analysis, restricting ourselves to a very rough and naive sketch. On the closure of each side s of $\partial\mathfrak{b}$ we define: $n_{\mathfrak{b};s}$, the unit outer normal to $\mathfrak{b}$ and on the edge of s, n_s the unit outer normal to s (n_s belongs to the tangent plane to $\mathfrak{b}$ and is normal to the edge). Then we assume that there exists a tensor function $G(x, n_{\mathfrak{b};s})$ on $(\partial\mathfrak{b} \times \mathcal{V})$ such that, on the edge e that belongs to the sides s_1, s_2, $f_{\mathfrak{b}}$ is given by

$$f_{\mathfrak{b}}|_e = G(x, n_{\mathfrak{b};s_1})n_{s_1} + G(x, n_{\mathfrak{b};s_2})n_{s_2}.$$

Thus, by the surface–divergence theorem, the last integral in (3.2.4) can be compacted within the second, surface, integral

$$\int_{\partial\mathfrak{b}} (t_{\mathfrak{b}} \cdot v + \operatorname{div}_{\partial\mathfrak{b}}(G^T v))d(\text{area}).$$

We conclude that, in particular, the following equation of balance of momentum must hold:

$$\int_{\mathfrak{b}} bd\,(\text{vol.}) \int_{\partial\mathfrak{b}} (t_{\mathfrak{b}} + \operatorname{div}_{\partial\mathfrak{b}} G)d(\text{area}) = 0,$$

an equation which coincides with (3.2.4) of [2].

3.4. Continua with Microstructure

The approach suggested in Section 3.2 can be adapted to deal with situations much more complex than the ones considered in Section 3.3. The essential step is the proof of a representation theorem leading to an expression of the working (3.2.1) more general than either (3.2.3) or (3.2.4).

Take the case of a body where each material element (the placement of which can be described with the usual tools) has, beside, a microstate which takes values ν in a compact manifold $\mathcal{M}$ of finite dimension m (see, e.g., [13]). Suppose $\mathcal{M}$ to be at least of class C^4 with a C^2–connection; then we can take advantage of results of Eliasson [14], as follows.

Let $\mathcal{B}$ and $\mathcal{B}_*$ be again the placements of such a body, call now $\bar{\mathcal{P}}$ the space of all *complete* placements: $\bar{\mathcal{P}} = C^0(\mathcal{B}_*; \mathcal{E} \times \mathcal{M})$, and consider associated fields of virtual velocities (v, υ); note that the microvelocity υ takes at y values $\upsilon(y)$ which belong to the tangent space $\mathcal{T}_{\nu(y)}\mathcal{M}$ of $\mathcal{M}$ at $\nu(y)$. Also, to make the following analysis more compact, introduce the notation

$$\bar{\nu} := (x, \nu) \qquad \bar{\mathcal{M}} := \mathcal{E} \times \mathcal{M} \qquad \bar{v} := (v, \upsilon).$$

Our goal is achieved through a version of the procedure proposed by Segev in [7] and [15]; our version does not involve at the outset the jet extension mapping. As we follow Segev's steps closely, we need only recall them here, properly adapted:

(i) Give $\bar{\mathcal{P}}$ the structure of a Banach manifold, after theorem 5.1 of [14], by virtue of our hypotheses on $\bar{\mathcal{M}}$.

(ii) For each complete placement $\bar{\nu} \in \bar{\mathcal{P}}$, identify the tangent space $\mathcal{T}_{\bar{\nu}}\bar{\mathcal{P}}$ with the continuous sections of the pullback $\bar{\nu}^*\tau_{\bar{\mathcal{M}}}$, i.e. of the vector bundle on $\mathcal{B}_*$, for which the fiber $\bar{\nu}^*\tau_{\bar{\mathcal{M}}}^{-1}$ at y is the tangent space $\mathcal{T}_{\bar{\nu}(y)}\bar{\mathcal{M}}$. In other words (see [14], Theorem 5.2 and the following Remark])

$$\begin{aligned}\mathcal{T}_{\bar{\nu}}\bar{\mathcal{P}} &\approx C^0(\bar{\nu}^*\tau_{\bar{\mathcal{M}}}) \approx C^0_{\bar{\nu}}(\mathcal{B}_*; \mathcal{T}\bar{\mathcal{M}})\\ &:= \{(\bar{\nu}, \bar{\upsilon}) \in C^0(\mathcal{B}_*; \mathcal{T}\bar{\mathcal{M}}) | (\bar{\nu}(y), \bar{\upsilon}(y)) \in \mathcal{T}_{\bar{\nu}(y)}\bar{\mathcal{M}}\ ,\ \forall y \in \mathcal{B}_*\}.\end{aligned}$$

$\mathcal{T}_{\bar{\nu}}\bar{\mathcal{P}}$ is therefore a banachable space (see [14], Sect. 4).

(iii) Given a field $\bar{\nu}$ find a measure representation of elements in $(\mathcal{T}_{\bar{\nu}}\bar{\mathcal{P}})^* = \mathcal{T}^*_{\bar{\nu}}\bar{\mathcal{P}}$. Such a representation is available when dealing with spaces of continuous functions defined on $\mathcal{B}_*$ and valued in a vector space, as we have recalled in Section 3.2. Here we are in a different setting; we can reach the goal, nevertheless, by trivializing locally the bundle $\bar{\nu}^*\tau_{\bar{\mathcal{M}}}$ and by splitting $C^0(\bar{\nu}^*\tau_{\bar{\mathcal{M}}})$ in a direct sum of finite number of spaces $C^0(\mathcal{B}_{*i}, \mathbf{R}^m)$ (see [16], pp. 10–11). The sets $\mathcal{B}_{*i}$ $(i = 1, 2, \ldots r)$ are regular subregions covering $\mathcal{B}_*$, such that, for any i, the values of $(\bar{\nu}(y), \bar{\upsilon}(y))$ $y \in \mathcal{B}_{*i}$ belong to one chart of $\mathcal{T}\bar{\mathcal{M}}$; thus we have the trivialization

$$\bar{\nu}^*\tau_{\bar{\mathcal{M}}}^{-1}(\mathcal{B}_{*i}) \to \mathcal{B}_{*i} \times \mathbf{R}^m,$$

and an associated partition of unity, so that we can identify $C^0(\bar{\nu}^*\tau_{\bar{\mathcal{M}}}|_{\mathcal{B}_{*i}})$ with $C^0(\mathcal{B}_{*i}, \mathbf{R}^m)$ and

$$\begin{aligned}C^0(\bar{\nu}^*\tau_{\bar{\mathcal{M}}}) &\approx \bigoplus_{i=1}^r C^0(\bar{\nu}^*\tau_{\bar{\mathcal{M}}}|_{\mathcal{B}_{*i}})\\ &\approx \Big\{(\omega_1, \ldots, \omega_r) \in \bigoplus_{i=1}^r C^0(\mathcal{B}_{*i}; \mathbf{R}^m) :\\ &\qquad \omega_i|_{\mathcal{B}_{*i}\cap\mathcal{B}_{*j}} = \omega_j|_{\mathcal{B}_{*i}\cap\mathcal{B}_{*j}}\Big\}\ .\end{aligned}$$

Now, for each $\pi^{ms}(\bar{\upsilon}) \in \mathcal{T}^*_{\bar{\nu}}\bar{\mathcal{P}} = [C^0(\bar{\nu}^*\tau_{\bar{\mathcal{M}}})]^*$ we can find $\pi_i^{ms}(\bar{\upsilon})$ $(i = 1, 2, \ldots r)$, $\pi_i^{ms} \in [C^0(\mathcal{B}_{*i}; \mathbf{R}^m)]^*$, such that

$$\pi^{ms}(\bar{\upsilon}) = \sum_{i=1}^r \pi_i^{ms}(\bar{\upsilon}), \qquad \forall \bar{\upsilon} \in \mathcal{T}_{\bar{\nu}}\bar{\mathcal{M}}$$

(see [7], Sect. IV). But $[C^0(\mathcal{B}_{*i}; \mathbf{R}^n)]^*$ is the space of Radon measures on $\mathcal{B}_{*i}$; hence we have finally

$$\pi^{m\sigma}(\bar{v}) = \sum_{i=1}^{r} \int_{\mathcal{B}_{*i}} \langle \bar{v}(y), d\bar{\mu}_{\mathcal{B}_{*i}}(y) \rangle = \int_{\mathcal{B}_{*}} \langle \bar{v}(y), d\bar{\mu}_{\mathcal{B}_{*}}(y) \rangle.$$

We can return now to earlier notation and separate components in the final result

$$\pi^{ms}(v, \upsilon) = \int_{\mathcal{B}_*} (v(y) \cdot dm_{\mathcal{B}_*}(y) + \langle \upsilon(y)\, ,\ d\mu_{\mathcal{B}_*} \rangle (y)). \qquad (3.4.1)$$

3.5. Balance Equations, When Microstructure Is Present

To exploit Noll's axiom we need to render explicit the effect of a change of observer on the reading of the value of the microstate, or, more particularly and specifically, the dependence of υ on the angular velocity w in a rigid motion.

We adopt suggestions and notation from [3], Sect. 3 and recall only the essential result expressing the rate of change of the microstate in a rigid motion

$$\upsilon^{(R)} = \mathcal{A} w;$$

here $\mathcal{A}$ is a linear operator from $\mathcal{V}$ into $\mathcal{T}_{\upsilon(y)}\mathcal{M}$: the infinitesimal generator of the group of rigid rotations on $\mathcal{M}$. We invoke Noll's axiom and require

$$\pi_{\mathfrak{b}}^{(ms)}(c + w \times x, \mathcal{A}w) = 0, \qquad \forall c, w \in \mathcal{V}.$$

Thus we arrive at the balance equations for momentum and moment of momentum

$$\int_{\mathfrak{b}} dm_{\mathfrak{b}} = 0, \qquad \int_{\mathfrak{b}} (x \times dm_{\mathfrak{b}} + \mathcal{A}^T d\mu_{\mathfrak{b}}) = 0, \qquad \forall \mathfrak{b} \in \mathcal{B}_*. \quad (3.5.1)$$

To go beyond we must again introduce some property of the measures; as for $\mu_{\mathfrak{b}}^{(1)}$ we may find convenient to accept the assumption that has led us to (3.2.3). Then $(3.5.1)_I$ implies again $(3.3.2)_I$. Similarly, we may assume that there exist $\beta(y)$ and $\sigma_{\mathfrak{b}}(y)$ in $\mathcal{T}^*_{\nu(y)}\mathcal{M}$ such that

$$\int_{\mathfrak{b}} \mathcal{A}^T d\mu_{\mathfrak{b}} = \int_{\mathfrak{b}} \mathcal{A}^T \beta d\,(\text{vol}) + \int_{\partial\mathfrak{b}} \mathcal{A}^T \sigma_{\mathfrak{b}} d\,(\text{area}),$$

with $\sigma_{\mathfrak{b}}(y)$ depending on $\mathfrak{b}$ only through the outer unit normal to $\partial\mathfrak{b}$ at y.

Then, instead of $(3.3.1)_{II}$, we obtain

$$\int_{\mathfrak{b}} (-\mathbf{e}T + \mathcal{A}^T \beta) d(\text{vol}) + \int_{\partial\mathfrak{b}} \mathcal{A}^T \sigma_{\mathfrak{b}} d(\text{area}) = 0,$$

where $\mathbf{e}$ is Ricci's permutation tensor. Following the standard proof of Cauchy's tetrahedron theorem we aressured first of the existence of a tensor S such that

$$\mathcal{A}^T \sigma_{\mathfrak{b}} = S n_{\mathfrak{b}}, \quad \text{on } \partial\mathfrak{b},$$

and then, by the divergence theorem and the usual technique of localization, of the validity of the local equation of balance of moment of momentum

$$\mathbf{e}T = \mathcal{A}^T \beta + \operatorname{div} S,$$

which takes here the place of $(3.3.2)_{II}$.

This result and even some the steps leading to it are substantially the same as can be found in [17], although the point of departure there is quite different. Further consequences derived there are the same as could be drawn here and, on these matters, we need only refer the reader to that monograph.

3.6. Acknowledgments

This work was carried out within the Research Groups of the Italian C.N.R., to which we belong.

We thank Professors M.K.V. Murthy and E. Virga for useful discussions.

References

[1] Noll, W. Lectures on the foundations of continuum mechanics and thermodynamics. *Arch. Rational Mech. Anal.*, **52** (1973), 62–92.

[2] Noll N. and Virga, E.G. On edge interactions and surface tension. *Arch. Rational Mech. Anal.*, **111** (1990), 1–31.

[3] Capriz, G. and Virga, E.G. Interactions in general continua with microstructure. *Arch. Rational Mech. Anal.*, **109** (1990), 323–342.

[4] Germain, P. Sur l'application de la méthode des puissances virtuelles en mécanique. *C. R. Acad. Sc. Paris*, **A 274** (1972), 1051–1055.

[5] Rougée, P. Formulation lagrangienne intrinsèque en mécanique des milieux continus. *J. Méch.*, **19** (1980), 7–32.

[6] Truesdell T. and Toupin, R.A. *The Classical Field Theories*. Encyclopedia of Physics, vol. III/1, Springer-Verlag, New York, (1960).

[7] Segev, R. Forces and the existence of stresses in invariant continuum mechanics. *J. Math. Phys.*, **27** (1986), 163–170.

[8] Serrin, J. The equations of continuum mechanics as a consequence of group invariance. Preprint, 1991.

[9] Noll, W. A new mathematical theory of simple materials. *Arch. Rational Mech. Anal.*, **48** (1972), 1–50.

[10] Dinculeanu, N. *Vector Measures*, Pergamon Press, Oxford, 1967.

[11] Forte, S. and Vianello, M. On surface stresses and edge forces. *Rend. Mat.*, **8** (1988), 409–426.

[12] Truesdell, C.A. *Rational Continuum Mechanics*, vol. I, 2nd edn. Academic Press, New York, 1991.

[13] Capriz, G. *Continua with Microstructure*, Springer-Verlag, New Yor, 1989.

[14] Eliasson, H.I. Geometry of manifolds of maps. *J. Differential Geom.*, **1** (1967), 169–194.

[15] Segev. R., A geometrical framework for the statics of materials with microsturcture. Preprint, 1992.
[16] Palais, R. *Foundations of Global Non-linear Analysis*. Benjamin, New York, 1968.
[17] Capriz, G. and Virga, E.G. On singular surfaces in the dynamics of continua with microstructure. *Quart. Appl. Math.*, **52** (1994), 509–517.

4
Entropy Numbers, Approximation Numbers, and Embeddings

D.E. Edmunds

4.1. Entropy Numbers

For convenience, all spaces considered here will be over the complex field. Given any Banach spaces X and Y, $L(X, Y)$ will stand for the space of all bounded linear maps from X to Y; we shall write $L(X)$ instead of $L(X, X)$. We also put $B_X = \{x \in X : \|x\| \leq 1\}$. Let $T \in L(X, Y)$ and $n \in \mathbb{N}$. Then the nth **entropy number** of T is defined to be

$$e_n(T) = \inf \left\{ \varepsilon > 0 : T(B_X) \subset \bigcup_{j=1}^{2^{n-1}} \{y_j + \varepsilon B_Y\} \text{ for some } y_1, \ldots, y_{2^{n-1}} \in Y \right\}.$$

A related concept can be found in the work of Kolmogorov in the 1930s: he considered any compact subset A of a metric space E, and defined the metric ε–entropy of A to be

$$H_\varepsilon(A) = \log_2 N_\varepsilon(A),$$

where $N_\varepsilon(A)$ is the least integer N such that A can be covered by N balls of radius ε. The connection between this and the entropy numbers is that, roughly speaking, $e_n(T)$ is the solution of the equation

$$H_\varepsilon(T(B_X)) = n - 1$$

for ε.

Various properties of entropy numbers can be established easily. Thus, if $S, T \in L(X, Y)$, $R \in L(Y, Z)$, and $m, n \in \mathbb{N}$, then

$$\|T\| = e_1(T) \geq e_2(T) \geq \cdots \geq 0,$$
$$e_{m+n-1}(S + T) \leq e_m(S) + e_n(T),$$
$$e_{m+n-1}(R \circ S) \leq e_m(R) e_n(S).$$

Moreover, if $\dim X = m < \infty$ and $I : X \to X$ is the identity map, then

$$1 \le 2^{(n-1)/(2m)} e_n(I) \le 4.$$

In addition, if $T \in L(X, Y)$, then as $n \to \infty$,

$$e_n(T) \downarrow \beta(T) = \inf\{\varepsilon > 0 : T(B_X) \text{ can be covered by finitely many balls of radius } \varepsilon\}.$$

Plainly $\beta(T) = 0$ if and only if T is compact; for this reason $\beta(T)$ is called the (ball) measure of noncompactness of T. It can be shown that

$$\tfrac{1}{2}\beta(T) \le \beta(T^*) \le 2\beta(T).$$

While in general these inequalities cannot be improved, nevertheless if X and Y are Hilbert spaces it turns out that $\beta(T) = \beta(T^*)$. There are important connections between $\beta(T)$ and spectral properties of T: thus if $T \in L(X)$, then

$$r_e(T) = \lim_{n\to\infty} \left(\beta(T^n)\right)^{1/n} \le \beta(T),$$

where $r_e(T)$ is the radius of the essential spectrum of T; and if $T \in L(H)$ is self-adjoint, H being a Hilbert space, then

$$r_e(T) = \beta(T).$$

For proofs of these assertions and much more we refer to [7], [10], [22], and [23].

The connection between the measure of noncompactness and spectral matters becomes even stronger if we consider certain embedding maps. Thus let Ω be an open subset of $\mathbb{R}^n$, let $W_2^1(\Omega)$ be the usual Sobolev space, and let $\mathring{W}_2^1(\Omega)$ be the closure of $C_0^\infty(\Omega)$ in $W_2^1(\Omega)$; let I, I_0 be the natural embeddings of $W_2^1(\Omega)$, $\mathring{W}_2^1(\Omega)$, respectively, in $L_2(\Omega)$. It emerges that $\beta(I_0) < 1$ if and only if the Friedrichs inequality holds; that is, there exists $c_0 > 0$ such that for all $u \in \mathring{W}_2^1(\Omega)$,

$$\int_\Omega |u(x)|^2\,dx \le c_0 \int_\Omega |\text{grad}\,u(x)|^2\,dx.$$

Furthermore, $\beta(I) = 0$ if and only if the spectrum of the Neumann Laplacian $\Delta_{N,\Omega}$ is discrete; if the Lebesgue n-measure $|\Omega|$ of Ω is infinite, then $\beta(I) = 1$; and if $|\Omega| < \infty$, then $\beta(I) < 1$ if and only if the Poincaré inequality holds, that is, there exists $c > 0$ such that for all $u \in W_2^1(\Omega)$,

$$\int \left| u(x) - |\Omega|^{-1} \int_\Omega u \right|^2 dx \le c \int_\Omega |\text{grad}\,u(x)|^2\,dx. \tag{4.1.1}$$

If $0 < \beta(I) < 1$, the spectrum of $\Delta_{N,\Omega}$ is not discrete, and the least point in the essential spectrum of $-\Delta_{N,\Omega}$ is $1/K$, where K is the infimum of the allowable constants c in 1. The celebrated "rooms and passages" domains provide examples in which $\beta(I)$ may take any value in [0, 1]. Details of the background and proofs of these results may be found in [1], [10], and [19], while further developments relating to the Friedrichs and Poincaré inequalities are contained in [11], [12], and [13].

The entropy numbers are also of considerable spectral importance. Let $T \in L(X)$ be compact. The spectrum of T, apart from the point 0, consists solely of eigenvalues of finite algebraic multiplicity: let $\{\lambda_k(T)\}_{k\in\mathbb{N}}$ be the sequence of all these eigenvalues, each repeated according to algebraic multiplicity and ordered by decreasing modulus, so that

$$|\lambda_1(T)| \geq |\lambda_2(T)| \geq \cdots \geq 0;$$

if T has only m ($< \infty$) distinct nonzero eigenvalues and M is the sum of their algebraic multiplicities, we adopt the convention that $\lambda_n(T) = 0$ if $n > M$. Probably the most useful connection between $\lambda_n(T)$ and $e_n(T)$ is that due to Carl [6]:

$$|\lambda_n(T)| \leq \sqrt{2}e_{n+1}(T) \qquad \text{for all} \quad n \in \mathbb{N}.$$

A more geometric proof of the stronger inequality

$$\left(\prod_{j=1}^{n} |\lambda_j(T)|\right)^{1/n} \leq 2^{(m-1)/(2n)} e_m(T) \qquad \text{for all} \quad m, n \in \mathbb{N}$$

is given by Carl and Triebel [8].

There are many open questions relating to the entropy numbers. For example, if $T \in L(H_1, H_2)$, where H_1 and H_2 are Hilbert space, then a polar decomposition argument shows that $e_n(T^*) = e_n(T)$ for all $n \in \mathbb{N}$, but outside Hilbert spaces the position is not at all clear: see Bourgain et al. [5] for recent work on this and numerous references. The behavior of the entropy numbers under interpolation is known if one endpoint space is held fixed (see Pietsch [23]) but nothing appears to be known when the spaces at both ends are allowed to vary. Finally, the entropy numbers of tensor products of maps have been studied to a certain extent (see Edmunds and Tylli [18], König [21], and Pietsch [24]) but much remains to be done.

4.2. Approximation Numbers

Given any $T \in L(X, Y)$ and any $n \in \mathbb{N}$, the nth approximation number of T is defined to be

$$a_n(T) = \inf\{\|T - A\| : \operatorname{rank} A < n\}.$$

Some of the properties of the $a_n(T)$ resemble those of the entropy numbers. For example, if $S, T \in L(X, Y)$, $R \in L(Y, Z)$ and $m, n \in \mathbb{N}$, then

$$\|T\| = a_1(T) \geq a_2(T) \geq \cdots \geq 0,$$
$$a_{m+n-1}(S+T) \leq a_m(S) + a_n(T),$$
$$a_{m+n-1}(R \circ S) \leq a_m(R)a_n(S).$$

However, there are radical differences: thus $a_m(T) = 0$ if and only if rank $T < m$; and if $\dim X \geq n$ and $I : X \to X$ is the identity map, then $a_n(I) = 1$. The behavior of the approximation numbers under the taking of adjoints appears to be better than that of the entropy numbers. For given any $T \in L(X, Y)$ and $n \in \mathbb{N}$, we have $a_n(T^*) \leq a_n(T) \leq 5a_n(T^*)$; and if T is compact, then $a_n(T^*) = a_n(T)$: see [10] and [17]. However, as will be apparent from the results about embedding maps given later, their behavior under interpolation is not good.

The most important connection between the approximation numbers and eigenvalues is that if H is a Hilbert space and $T \in L(H)$ is compact, then $a_n(T) = \lambda_n(|T|)$ for all $n \in \mathbb{N}$, where $|T|$ is the positive square root of T^*T. This implies that if, in addition, T is positive and self-adjoint, then its approximation numbers and eigenvalues coincide.

It is natural to seek to compare the entropy and approximation numbers of a map $T \in L(X, Y)$. General results are available which imply, for example, that if $a_n(T) = O(n^{-\rho})$ for some $\rho > 0$ (as $n \to \infty$), then the same holds for $e_n(T)$; a similar result holds for decay like some negative power of $\log n$. Details of this may be found in Triebel [29]. Despite this, however, we shall see later that the behavior of $a_n(T)$ may be quite different from that of $e_n(T)$ for certain embedding maps T.

4.3. Embeddings

These are important because the maps which turn up in connection with differential or integral operators can often be factorized into the composition of a continuous map and an embedding between function spaces. We deal with them in the context of the scales of spaces $B^s_{p,q}$ and $F^s_{p,q}$.

To define these, let ϕ be a function in the Schwartz space $S(\mathbb{R}^n)$ such that $\operatorname{supp}\phi \subset \{y \in \mathbb{R}^n : |y| < 2\}$, $\phi(x) = 1$ if $|x| \leq 1$, and put $\phi_j(x) = \phi(2^{-j}x) - \phi(2^{-j+1}x)$ for all $j \in \mathbb{N}$, $\phi_0 = \phi$; note that for all $x \in \mathbb{R}^n$, $1 = \sum_{j=0}^{\infty} \phi_j(x)$, and that $(\phi_j \hat{f})^\vee$ is an entire analytic function on $\mathbb{R}^n$ ($\hat{f}$ standing for the Fourier transform of f and $\vee$ denoting the inverse Fourier transform). Given any $s \in \mathbb{R}$, $p \in (0, \infty]$ and $q \in (0, \infty]$, we define

$$B^s_{p,q}(\mathbb{R}^n) = \left\{ f \in S'(\mathbb{R}^n) : \|f \,|B^s_{p,q}(\mathbb{R}^n)\| = \left(\sum_{j=0}^{\infty} 2^{jsq} \|(\phi_j \hat{f})^\vee | L_p\|^q \right)^{1/q} < \infty \right\}$$

for $q < \infty$, with the obvious modification if $q = \infty$. Correspondingly, if $s \in \mathbb{R}$, $p \in (0, \infty)$ and $q \in (0, \infty]$ we define

$$F^s_{p,q}\left(\mathbb{R}^n\right) = \left\{ f \in S'\left(\mathbb{R}^n\right) : \left\| \left(\sum_{j=0}^{\infty} 2^{jsq} \left| \left(\phi_j \hat{f}\right)^\vee \right|^q \right)^{1/q} \Big| L_p \right\| < \infty \right\}$$

for $q < \infty$, again with the obvious modification if $q = \infty$. It can be shown that for the given parameter values, $B^s_{p,q}\left(\mathbb{R}^n\right)$ and $F^s_{p,q}\left(\mathbb{R}^n\right)$ are quasi-Banach spaces (that is, having the same properties as Banach spaces save that the triangle inequality is replaced by the inequality $\|f + g\| \le c\left(\|f\| + \|g\|\right)$ for some $c \ge 1$), and that they are Banach spaces if $p, q \ge 1$; moreover, they are independent (up to equivalence of (quasi) norms) of the particular choice of the generating function ϕ.

Many well-known spaces are included in these scales. Thus if $s > 0$, $B^s_{\infty,\infty}\left(\mathbb{R}^n\right)$ is the Hölder–Zygmund space; if $s \in \mathbb{R}$, $1 < p < \infty$, $1 \le q \le \infty$, then $B^s_{p,q}\left(\mathbb{R}^n\right)$ is the classical Besov space; if $s \in \mathbb{R}$, $1 < p < \infty$, then $F^s_{p,2}\left(\mathbb{R}^n\right)$ is the fractional Sobolev space (or Bessel potential space) $H^s_p\left(\mathbb{R}^n\right)$, becoming the classical Sobolev space $W^s_p\left(\mathbb{R}^n\right)$ if $s \in \mathbb{N}$ and the Lebesgue space $L_p\left(\mathbb{R}^n\right)$ if $s = 0$; and if $0 < p < \infty$, then $F^s_{p,2}\left(\mathbb{R}^n\right)$ is the nonhomogeneous Hardy space $h_p\left(\mathbb{R}^n\right)$, coinciding with $L_p\left(\mathbb{R}^n\right)$ if $1 < p < \infty$.

Now let Ω be a bounded domain in $\mathbb{R}^n$ with C^∞ boundary. The spaces $B^s_{p,q}(\Omega)$ and $F^s_{p,q}(\Omega)$ are defined by restriction to Ω. Thus

$$B^s_{p,q}(\Omega) = \left\{ f \in \mathcal{D}'(\Omega) : \text{ there exists } g \in B^s_{p,q}\left(\mathbb{R}^n\right) \text{ with } g|_\Omega = f \right\}$$

and

$$\left\| f \,|B^s_{p,q}(\Omega) \right\| = \inf \left\{ \left\| g \,|B^s_{p,q}\left(\mathbb{R}^n\right) \right\| : g|_\Omega = f \right\};$$

$F^s_{p,q}(\Omega)$ is similarly defined.

For details of these spaces we refer to the comprehensive account given by Triebel [26] and [27].

We now turn to entropy and approximation numbers of embeddings between these spaces, and note that the definition of these numbers can be taken over without change to a quasi-Banach space setting, with only slight modification of their properties resulting from the weakening of the triangle inequality. For convenience, we shall let $A^s_{p,q}(\Omega)$ stand for either $B^s_{p,q}(\Omega)$ or $F^s_{p,q}(\Omega)$, with the understanding that for $F^s_{p,q}(\Omega)$ we must have $p < \infty$. The basic result is the following, due to Edmunds and Triebel [14], [15]:

Let Ω be a bounded domain in $\mathbb{R}^n$ with C^∞ boundary, let $s_1, s_2 \in \mathbb{R}$; $p_1, p_2, q_1, q_2 \in (0, \infty]$; and suppose that

$$\delta := s_1 - s_2 - n\left(\frac{1}{p_1} - \frac{1}{p_2}\right)_+ > 0.$$

Then the natural embedding $I : A^{s_1}_{p_1,q_1}(\Omega) \to A^{s_2}_{p_2,q_2}(\Omega)$ exists and is compact, and

$$e_k(I) \asymp k^{-(s_1-s_2)/n},$$
$$a_k(I) = O\left(k^{-\delta/n}\right).$$

More precisely,

$$a_k(I) \asymp k^{-\delta/n} \qquad \text{if} \quad 0 < p_1 \le p_2 \le 2 \quad \text{or} \quad 2 \le p_1 \le p_2 \le \infty$$
$$\text{or} \quad 0 < p_2 \le p_1 \le \infty;$$

$$a_k(I) \ge ck^{-\delta/n-\varepsilon} \qquad \text{if} \quad 0 < p_1 \le 2 \le p_2 \le \infty,$$

where

$$\varepsilon = \min\left(\frac{1}{p_1} - \frac{1}{2}, \frac{1}{2} - \frac{1}{p_2}\right);$$

and in some cases, $k^{-\delta/n-\varepsilon}$ gives the correct asymptotic order of $a_k(I)$, not $k^{-\delta/n}$.

The complicated behavior of the approximation numbers shows that the unrestricted interpolation property cannot hold for them.

Results of this character go back, at least for classical Sobolev spaces, to the work of Birman and Solomjak [2], [3], [4], who used a technique based on piecewise polynomial approximation. The methods used to prove the above theorem are those of Fourier analysis, and so far as the approximation numbers are concerned also involve the reduction of the problem to that of embeddings of the form

$$I_l : l_{p_1}^{2^{jn}} \to l_{p_2}^{2^{jn}}$$

for which precise estimates for the approximation numbers are available.

To conclude this section, we mention some recent work on embeddings in Orlicz spaces. Let Ω be a bounded domain in $\mathbb{R}^n$ with smooth boundary, and let $1 < p < \infty$, $m \in \mathbb{N}$, $mp = n$, $1 < \nu < p'$; put $\phi_\nu(t) = \exp(t^\nu) - 1$ $(t \ge 0)$. The Orlicz space $L_{\phi_\nu}(\Omega)$ is the liner hull of all those measurable functions u on Ω such that $\int_\Omega \phi_\nu(|u(x)|)\,dx < \infty$; endowed with the Luxemburg norm

$$\|u\,|L_{\phi_\nu}(\Omega)\,\| = \inf\left\{\lambda > 0 : \int_\Omega \phi_\nu(|u(x)|/\lambda)\,dx \le 1\right\}$$

it is a Banach space. In this case, with $mp = n$, it is of course known that $W_p^m(\Omega)$ is continuously embedded in every $L_q(\Omega)$ space, but Trudinger [30] proved that $W_p^1(\Omega)$ is compactly embedded in $L_{\phi_\nu}(\Omega)$; the same was proved for $H_p^{n/p}(\Omega)$ $(1 < p < \infty,\ n \in \mathbb{N})$ by Strichartz [25] and for $B_{p,p}^{n/p}(\Omega)$ $(1 < p < \infty,\ n \in \mathbb{N})$ in [9]. Denoting by I any one of these embeddings it was known [9] that

$$a_k(I), e_k(I) = O\left((\log k)^{1/p'-1/\nu}\right);$$

but Triebel [28] has recently shown that in fact

$$a_k(I) \asymp (\log k)^{1/p'-1/\nu},$$

and for all $\varepsilon > 0$, there exist constants $c_1, c_2(\varepsilon) > 0$ such that

$$c_1 k^{-(1/\nu-1/p')} \le e_k(I) \le c_2(\varepsilon) k^{-1/3(1/\nu-1/p')+\varepsilon}.$$

If $1/\nu > 1 + 2/p$, then $e_k(I) \asymp k^{-1/p}$. Some sharpening of the entropy number estimates seems possible.

4.4. Applications

The principal application to date of the entropy numbers is to combine Carl's inequality, relating eigenvalues to entropy numbers, with the sharp estimates mentioned above for the entropy numbers of embedding maps and so to obtain estimates for the eigenvalues of (degenerate) elliptic (pseudo-) differential operators with nonsmooth coefficients and their inverses. Details of the results that can be obtained by this means, in the context of a bounded domain Ω in $\mathbb{R}^n$, are contained in [16], which also contains a description of the papers by Haroske and Triebel [20] which extend this to $\mathbb{R}^n$, having first obtained entropy number estimates for embeddings of weighted function spaces.

References

[1] Amick, C.J. Some remarks on Rellich's theorem and the Poincaré inequality. *J. London Math. Soc.* **18** (1978), 81–93.

[2] Birman, M.S. and Solomjak, M.Z. Spectral asymptotics of non-smooth elliptic operators, I. *Trans. Moscow Math. Soc.*, **27** (1972), 1–52.

[3] Birman, M.S. and Solomjak, M.Z. Spectral asymptotics of non-smooth elliptic operators, II. *Trans. Moscow Math. Soc.*, **28** (1973), 1–32.

[4] Birman, M.S. and Solomjak, M.Z. Quantitative analysis in Sobolev imbedding theorems and applications to spectral theory. *Amer. Math. Soc. Transl.*, **114** (1980), 1–132.

[5] Bourgain, J., Pajor, A., Szarek, S.J., and Tomczak-Jaegermann, N. On the duality problem for entropy numbers of operators. In *Geometric Aspects of Functional Analysis.* Lecture Notes in Mathematics, Vol. 1376. Springer-Verlag, New York, 1989.

[6] Carl, B. Entropy numbers, s-numbers and eigenvalue problems. *J. Funct. Anal.*, **41** (1981), 290–306.

[7] Carl, B. and Stephani, I. *Entropy, Compactness and the Approximation of Operators.* Cambridge University Press, Cambridge, 1990.

[8] Carl, B. and Triebel, H. Inequalities between eigenvalues, entropy numbers and related quantities in Banach spaces. *Math. Ann.*, **251** (1980), 129–133.

[9] Edmunds, D.E., Edmunds, R.M., and Triebel, H. Entropy numbers of embeddings of fractional Besov–Sobolev spaces in Orlicz spaces. *J. London Math. Soc.*, **35** (1987), 121–134.

[10] Edmunds, D.E. and Evans, W.D. *Spectral Theory and Differential Operators.* Oxford University Press, Oxford, 1987.

[11] Edmunds, D.E. and Opic, B. Weighted Poincaré and Friedrichs inequalities. *J. London Math. Soc.* **47** (1993), 79–93.
[12] Edmunds, D.E., Opic, B., and Pick, L. Poincaré and Friedrichs inequalities in abstract Sobolev spaces. *Math. Proc. Cambridge Philos. Soc.*, **113** (1993), 355–379.
[13] Edmunds, D.E., Opic, B., and Rákosník, J. Poincaré and Friedrichs inequalities in abstract Sobolev spaces, II. *Math. Proc. Cambridge Philos. Soc.*, **115** (1994), 159–173.
[14] Edmunds, D.E. and Triebel, H. Entropy numbers and approximation numbers in function spaces. *Proc. London Math. Soc.*, **58** (1989), 137–152.
[15] Edmunds, D.E. and Triebel, H. Entropy numbers and approximation numbers in function spaces, II. *Proc. London Math. Soc.*, **64** (1992), 153–169.
[16] Edmunds, D.E. and Triebel, H. *Function Spaces, Entropy Numbers, Differential Operators*. Cambridge University Press, Cambridge, 1996.
[17] Edmunds, D.E. and Tylli, H.-O. On the entropy numbers of an operator and its adjoint. *Math. Nachr.*, **126** (1986), 231–239.
[18] Edmunds, D.E. and Tylli, H.-O. Entropy numbers of tensor products of operators. *Ark. Mat.*, **31** (1993), 247–274.
[19] Evans, W.D. and Harris, D.J. Sobolev embeddings for ridged domains. *Proc. London Math. Soc.*, **54** (1987), 141–176.
[20] Haroske, D., and Triebel, H. Entropy numbers in weighted function spaces and eigenvalue distributions of some degenerate pseudodifferential operators, I. *Math. Nachr.*, **167** (1994), 131–156; II, *ibid*, **168** (1994), 109-137.
[21] König, H. On the tensor stability of s-number ideals. *Math. Ann.*, **269** (1984), 77–93.
[22] König, H. *Eigenvalue Distribution of Compact Operators*. Birkhäuser, Basel, 1986).
[23] Pietsch, A. *Operator Ideals*. North-Holland, Amsterdam, 1980.
[24] Pietsch, A. Tensor products of sequences, functions and operators. *Arch. Math.*, **38** (1982), 335–344.
[25] Strichartz, R.S. A note on Trudinger's extension of Sobolev's inequality. *Indiana Univ. Math. J.*, **21** (1972), 841–842.
[26] Triebel, H. *Theory of Function Spaces*. Birkhäuser, Basel, 1983.
[27] Triebel, H. *Theory of Function Spaces*, II. Birkhäuser, Basel, 1992.
[28] Triebel, H. Approximation numbers and entropy numbers of embeddings of fractional Besov–Sobolev spaces in Orlicz spaces. *Proc. London Math. Soc.*, **66** (1993), 589–618.
[29] Triebel, H. Relations between approximation numbers and entropy numbers. *J. Approx. Theory*, **78** (1994), 112–116.
[30] Trudinger, N.S. On imbedding into Orlicz spaces and some applications. *J. Math. Mech.*, **17** (1967), 473–484.

5
Some Regularity Properties of Locally Weakly Invertible Maps

M. Giaquinta, G. Modica, and J. Souček

5.1. Introduction

Let Ω be an open bounded set in $\mathbf{R}^n$, and let us denote by $\hat{\mathbf{R}}^n$ just another copy of $\mathbf{R}^n$ and let u be a function in $W^{1,1}(\Omega, \hat{\mathbf{R}}^n)$. We denote by $\mathcal{L}_u$, $\mathcal{L}_{Du}$ the Lebesgue sets of u and Du and with $\overline{u}$ a Lebesgue representative of u, i.e.,

$$\overline{u}(x) := \begin{cases} \text{Lebesgue value of } u \text{ at } x & \text{if } x \in \mathcal{L}_u, \\ \text{constant} & \text{if } x \notin \mathcal{L}_u, \end{cases}$$

and by $\tilde{u} : \Omega \to \hat{\mathbf{R}}^n$ the "restriction" of $\overline{u}$ to the regular points of u, i.e.,

$$\tilde{u}(x) := \begin{cases} \text{Lebesgue value of } \overline{u}(x) & \text{if } x \in \mathcal{R}_u, \\ \text{constant} & \text{if } x \notin \mathcal{R}_u, \end{cases}$$

where $\mathcal{R}_u := \mathcal{L}_u \cap \mathcal{L}_{Du}$. Recall that $\tilde{u}$ is approximately differentiable on $\mathcal{R}_u$ and ap $D\tilde{u} = Du$ a.e. From now on we always denote by $D\tilde{u}(x)$ the approximate differential ap$D\tilde{u}(x)$. Recall also that $\tilde{u}$ *is a Lusin representative of* u since $\tilde{u}$ satisfy the Lusin property (N): $|\tilde{u}(A)| = 0$ if $|A| = 0$ and moreover the area formula says that

$$\int_A |\det Du(x)|\, dx = \int_{\mathbf{R}^n} N(u, A, y)\, dy, \tag{5.1.1}$$

where

$$N(u, A, y) = N(\tilde{u}, A, y) = \#\{x \mid x \in A \cap \mathcal{R}_u,\ \tilde{u}(x) = y\} \tag{5.1.2}$$

is the Banach indicatrix of $\tilde{u}$ and it is easy to see that $N(u, A, \cdot)$ is independent of the chosen Lusin's representative in the L^1 sense.

Let us denote by $M(Du)$ the simple n-vector in $\Lambda_n(\mathbf{R}^n \times \hat{\mathbf{R}}^n)$ given by

$$M(Du) := (e_1 + D_1 u) \wedge (e_2 + D_2 u) \wedge \cdots \wedge (e_n + D_n u),$$

where $e_1, \dots, e_n$ denotes the canonical basis of $\hat{\mathbf{R}}^n$. If $u \in W^{1,1}(\Omega, \hat{\mathbf{R}}^n)$ and $|M(Du)| \in L^1(\Omega)$, the *graph current* associated with u is the n-dimensional rectifiable current

$$G_u = \tau(\mathcal{M}, 1, \xi),$$

where $\mathcal{M} := \{(x, \tilde{u}(x)) \,|\, x \in \mathcal{R}_u\}$ and $\xi(x, \tilde{u}(x))$ is the n-vector orienting $\mathcal{M}$

$$\xi(x, \tilde{u}(x)) := \frac{M(Du(x))}{|M(Du(x))|}.$$

The area formula then says that

$$\mathbf{M}(G_u) = \mathcal{H}^n(\mathcal{G}_u) = \int_\Omega |M(Du)|\, dx.$$

The class of *Cartesian maps* is defined by

$$\operatorname{cart}^1(\Omega, \hat{\mathbf{R}}^n) := \{u \in W^{1,1}(\Omega, \hat{\mathbf{R}}^n) \,|\, |M(Du)| \in L^1(\Omega),\ \partial G_u \llcorner \Omega \times \hat{\mathbf{R}}^n = 0\}$$

and the classes of *local and global weak diffeomorphisms* by

$$\operatorname{cart}^1_+(\Omega, \hat{\mathbf{R}}^n) := \{u \in \operatorname{cart}^1(\Omega, \hat{\mathbf{R}}^n) \,|\, \det Du > 0 \text{ a.e. in } \Omega\},$$
$$\widetilde{\operatorname{dif}}^{1,1}(\Omega, \hat{\mathbf{R}}^n) := \{u \in \operatorname{cart}^1(\Omega, \hat{\mathbf{R}}^n) \,|\, N(u, \Omega, y) \le 1 \text{ a.e. in } \hat{\mathbf{R}}^n\};$$

compare [5], [4], and [10].

Motivated by physical considerations, we gave in [10] a weak formulation of the kinematics of finite elasticity. According to [10], and replacing the physical dimension 3 by n, an *elastic deformation* is a map $u : \Omega \to \hat{\mathbf{R}}^n$, Ω being an open and connect subset of $\mathbf{R}^n$ and $\hat{\mathbf{R}}^n$ a copy of $\mathbf{R}^n$, such that:

(i) u has an almost everywhere inverse $\hat{u}$;

(ii) u and $\hat{u}$ satisfy Lusin property (N)

$$|u(A)| = 0 \qquad \text{iff} \quad |A| = 0;$$

(iii) u and, consequently, $\hat{u}$ are almost everywhere approximately differentiable, $\det Du \ge 0$ a.e. in Ω; and

(iv) the integer rectifiable current G_u carried by the graph of u is boundaryless in $\Omega \times \hat{\mathbf{R}}^n$.

We proved also in [10] that the class of elastic deformations coincides with the class of *weak diffeomorphisms*, $\widetilde{\operatorname{dif}}^{1,1}(\Omega, \hat{\mathbf{R}}^n)$, introduced in [5], compare also [4].

In this chapter we will discuss some pointwise properties of maps in $\operatorname{cart}^1_+(\Omega, \hat{\mathbf{R}}^n) \cap W^{1,p}(\Omega, \hat{\mathbf{R}}^n)$, $p > n-1$. Similar properties have been proved for maps in the subclass $\mathcal{A}_{p,q}$, $p \ge n-1$, $q > n/n-1$, in [24], [21], and [23], i.e., assuming higher integrability of the adjoint matrix adjDu. It is one of the aims of this chapter to show that higher integrability of adjDu in not necessary, and that those properties depend only on the condition

$$\partial G_u \llcorner \Omega \times \hat{\mathbf{R}}^n = 0, \tag{5.1.3}$$

which for maps in $W^{1,n-1}(\Omega, \hat{\mathbf{R}}^n)$ just amounts to the following formula of integration by parts

$$\int_\Omega \sum_{i,j} D_j[\phi(x, u(x))](\mathrm{adj} Du)^i_j \, dx = 0$$

for all $i = 1, 2, \ldots, n$ and all $\phi \in C_c^\infty(\Omega \times \hat{\mathbf{R}}^n)$.

In Section 5.4 we will also prove that the range, $\tilde{u}(\Omega)$ of any map $u \in \mathrm{cart}^1_+(\Omega, \hat{\mathbf{R}}^n) \cap W^{1,p}(\Omega, \hat{\mathbf{R}}^n)$, $p > n-1$, is a.e. *open*, in particular every weak diffeomorphism $u \in \widetilde{\mathrm{dif}}^{1,1}(\Omega, \hat{\mathbf{R}}^n) \cap W^{1,p}(\Omega, \hat{\mathbf{R}}^n)$, $p > n-1$, defines an a.e. homeomorphism between Ω and $\tilde{u}(\Omega)$.

We would like to emphasize that condition (5.1.3) is relevant in many questions. It prevents fractures, as for instance cavitation, in finite elasticity, i.e., in defining an elastic deformation, compare [5], [10], and [11]. It is a necessary condition for the weak approximability by smooth maps, [5], though in general not a sufficient one, compare [19] and [12]. It is the "right" condition in ensuring weak convergence of Jacobian determinants, [5], [4], [20], and [3], which fails if (5.1.3) does not hold, compare [1]. It allows us to discuss in a reasonable way topological properties such as degree [4], [13] and connectivity of the range [11], or to define the homological map u_* [8], [9] between homology groups. Finally, it is the key point in approaching the problem of harmonic maps into manifolds $\mathcal{Y}$ with nontrivial topology, [4], [6], in particular, in understanding the continuity properties of the **D**-field in [2], and in discussing the Oseen–Frank model for liquid crystals [7].

5.2. Local Weak Diffeomorphisms

Let Ω be an open bounded set in $\mathbf{R}^n$, let us denote by $\hat{\mathbf{R}}^n$ just another copy of $\mathbf{R}^n$, and let u be a function in $W^{1,1}(\Omega, \hat{\mathbf{R}}^n)$. It is convenient to denote by $\mathcal{A}_u$ the set of points in Ω where u is approximately continuous. Of course, $\mathcal{R}_u \subset \mathcal{L}_u \subset \mathcal{A}_u$.

The degree of $u \in W^{1,1}(\Omega, \hat{\mathbf{R}}^n)$ is defined by

$$\deg(u, \Omega, y) := \sum_{x \in \tilde{u}^{-1}(y)} \mathrm{sgn}(\det D\tilde{u}(x)), \tag{5.2.1}$$

where the sum in (5.2.1) is extended only to points $x \in \mathcal{R}_u$. The degree is independent in the L^1 sense of the chosen representative and, by the area formula, we have for $u \in W^{1,1}(\Omega, \hat{\mathbf{R}}^n)$ that $\det Du \in L^1(\Omega)$ if and only if $\deg(u, \Omega, y) \in L^1(\hat{\mathbf{R}}^n)$ and

$$\int_\Omega \det Du\, dx = \int_{\hat{\mathbf{R}}^n} \deg(u, \Omega, y)\, dy.$$

We can show, compare, e.g., [13], that the degree satisfies the usual properties of the degree. In particular, for $u \in \mathrm{cart}^1_+(\Omega, \hat{\mathbf{R}}^n)$ we have

$$N(u, A, y) = \deg(u, A, y) \qquad \text{a.e.} \quad y \in \hat{\mathbf{R}}^n.$$

The first claim below is just a consequence of degree theory, compare [13]. For any subset A of Ω and $u \in \mathrm{cart}^p_+(\Omega, \widehat{\mathbf{R}}^n)$, let us denote by

$$\Gamma_{G_u, A} := \widehat{\mathbf{R}}^n \setminus \mathrm{spt} \widehat{\pi}_\# \,\partial(G_u \llcorner A \times \widehat{\mathbf{R}}^n), \tag{5.2.2}$$

where $\widehat{\pi} : \mathbf{R}^n \times \hat{\mathbf{R}}^n \to \hat{\mathbf{R}}^n$ is the projection on the second factor, $\pi(x, y) = y$.

Theorem 5.2.1. *Let A be a subset of Ω and $u \in \mathrm{cart}^p_+(\Omega, \widehat{\mathbf{R}}^n)$, $p \geq 1$. Also, let $\{W_k\}$ be the decomposition of the open set $\Gamma_{G_u, A}$ into connected components. Then there exist nonnegative integers m_k, one for each W_k such that:*

(i) $\deg(u, A, y) = m_k$ *for a.e.* $y \in W_k$.

(ii) $m_k = 0$ *if* $|W_k| = \infty$.

(iii) $N(u, \Omega, y) = m_k$ *for a.e.* $y \in W_k$*, i.e., a.e.* $y \in W_k$ *has exactly k preimages (in* $\mathcal{R}_u$*).*

(iv) *Suppose that A has density one at each of its points, $A \subset (A)_1$. If there exists $x \in A \cap \mathcal{A}_u$ such that $\overline{u}(x) \in W_k$ for some k, then $m_k \geq 1$. Or, in other words, if $m_k = 0$, then*

$$\overline{u}(A \cap \mathcal{A}_u) \cap W_k = \emptyset.$$

(v) *$\tilde{u}$ has the double Lusin's property* (N)

$$|A \cap \mathcal{R}_u| = 0 \qquad \textit{iff} \qquad |\tilde{u}(A \cap \mathcal{R}_u)| = 0.$$

(vi) *Let $a \in \Omega$. Then for a.e., r, $0 < r < \mathrm{dist}(a, \partial\Omega)$, we have*

$$\partial(G_u \llcorner B(a, r) \times \widehat{\mathbf{R}}^n) = G_{\tilde{u} \,|\, \partial B(a,r)},$$

where $G_{\tilde{u} \,|\, \partial B(a,r)}$ denotes the current integration over the graph of the trace of u on $\partial B(a, r)$. In particular, for a.e. r we have

$$\int_{\mathbf{R}^n} \operatorname{div} g(y) \deg(u, B(0, r), y)\, dy =$$
$$\int_{\partial B(0,r)} \sum_{i,j=1}^n g^i(u(x))\, (\mathrm{adj} Du(x))^i_j\, \nu_j(x)\, d\mathcal{H}^{n-1}$$

for all $g \in C^1_c(\mathbf{R}^n, \mathbf{R}^n)$. *Therefore*

$$\mathcal{H}^{n-1}(\partial B(a,r)\backslash \mathcal{R}_u) = 0,$$
$$\operatorname{spt}\widehat{\pi}_\# \partial(G_u \llcorner B(a,r) \times \widehat{\mathbf{R}}^n) \subset \overline{\tilde{u}(\partial B(a,r) \cap \mathcal{R}_u)}.$$

Proof. For (i), (ii), and (vi), compare [13]. Part (iii) follows from (ii), and (v) follows from the area formula since $\det Du > 0$ a.e. We only have to prove (iv). Let $x \in A$, and $\overline{u}(x) \in W_k$. Let B be a ball centered at $\overline{u}(x)$ and inside W_k. By definition of $\overline{u}(x)$, $\theta(u^{-1}(B), x) = 1$ and since $A \subset (A)_1$, $\theta(u^{-1}(B) \cap A, x) = 1$. If now $C := u^{-1}(B) \cap A$, then $|C| > 0$ and

$$0 < \int_C \det Du\, dx = \int N(u, C, y)\, dy.$$

Hence $|\tilde{u}(C)| > 0$, $\tilde{u}(C) \subset W_k$ and $N(u, C, y) \geq 1$ on $\tilde{u}(C)$. This excludes that

$$\deg(u, \Omega, y) = N(u, \Omega, y) = 0 \qquad \text{a.e.} \quad y \in W_k$$

and then $m_k \neq 0$.

As a simple corollary we get:

Corollary 5.2.1. *Let* $u \in \operatorname{cart}^1_+(\Omega, \widehat{\mathbf{R}}^n)$. *Suppose that there exists a bounded open simply connected domain* $\widehat{\Omega} \subset \widehat{\mathbf{R}}^n$ *such that*

$$\operatorname{spt}\partial G_u \subset \partial\Omega \times \partial\widehat{\Omega}.$$

Then there exists a positive integer $m > 0$ *such that* $N(u, \Omega, y) = m$ *for a.e.* $y \in \widehat{\Omega}$. *Moreover,*

$$\overline{u}(\Omega \cap \mathcal{A}_u) \subset \overline{\widehat{\Omega}}.$$

In particular, if $|\partial\widehat{\Omega}| = 0$, *then the Lebesgue representative* $\overline{u}$ *of* u *has the Lusin property* (N) *too.*

5.3. Regularity Properties of Maps in $\operatorname{cart}^1_+ \cap W^{1,p}$, $p > n-1$

Let Ω be a bounded and connected open set in $\mathbf{R}^n$, and let $u \in W^{1,p}(\Omega, \hat{\mathbf{R}}^n)$ for some $p > n-1$. From [14] and [25] we know that the Hausdorff dimension of $\Omega\backslash\mathcal{L}_u$ is not larger than $n-p$, hence

$$\dim_{\mathcal{H}}(\Omega\backslash\mathcal{L}_u) < 1.$$

Moreover, for a.e. r we can assume that the trace of u on $\partial B(a,r)$ belongs to $W^{1,p}(\partial B(a,r), \hat{\mathbf{R}}^n)$ and by the Morrey–Sobolev embedding theorem that it has a continuous representative $\overline{u}$ on $\partial B(a,r)$ which agrees with the Lebesgue representative of u on the whole of $\partial B(a,r)$. More precisely, we can state:

Lemma 5.3.1. *Let $u \in W^{1,p}(\Omega, \widehat{\mathbf{R}}^n)$, $p > n-1$, and let $a \in \Omega$. Then for almost every r with $0 < r < r_a := \operatorname{dist}(a, \partial\Omega)$ we have:*

(i) $\partial B(a,r) \subset \mathcal{L}_u$.

(ii) *The trace of u on $\partial B(a,r)$ agrees with the approximately continuous representative $\overline{u}$ of u. Moreover, $\overline{u}$ is Hölder continuous on $\partial B(a,r)$ and belongs to $W^{1,p}(\partial B(a,r), \hat{\mathbf{R}}^n)$ and*

$$\mathcal{H}^{n-1}(\overline{u}(A)) = 0 \quad \textit{provided} \quad A \subset \partial B(a,r) \quad \text{and} \quad \mathcal{H}^{n-1}(A) = 0. \tag{5.3.1}$$

(iii) *We also have*

$$\operatorname{osc}(\overline{u}, \partial B(a,r)) \leq c\, r^{1-(n-1)/p} \left(\int_{\partial B(a,r) \cap \mathcal{R}_u} |D\tilde{u}|^p \, d\mathcal{H}^{n-1} \right)^{1/p},$$

where $c = c(n,p)$ and for $A \subset \Omega$

$$\operatorname{osc}(\overline{u}, A) := \sup\{ |\overline{u}(x) - \overline{u}(y)| \mid x, y \in A \cap \mathcal{A}_u \}.$$

Proof. For the reader's convenience we just prove (5.3.1), compare, e.g., [22]. Fix $\varepsilon > 0$ and let $\{B(x_k, r_k)\}$ be a covering by balls of A with $x_k \in \partial B(a, r_a)$ and such that

$$\sum_k r_k^{n-1} < \varepsilon.$$

Denote by $d_k := \operatorname{diam}(\overline{u}(B(x_k, r_k) \cap \partial B(a, r_a)))$. By Morrey's embedding theorem,

$$d_k \leq c r_k^{1-(n-1)/p} \left(\int_{B(x_k,r_k) \cap \partial B(a,r_a)} |Du|^p \, d\mathcal{H}^{n-1} \right)^{(n-1)/p}.$$

Since the balls $\{B(\overline{x_k}, d_k)\}$ form a covering of $\overline{u}(A)$, we get, using the Hölder inequality,

$$\mathcal{H}^{n-1}(\overline{u}(A)) \leq \sum_k d_k^{n-1} \leq c \left(\sum_k r_k^{n-1} \right)^{1-(n-1)/p}$$
$$\left(\sum_k \int_{B(x_k,r_k) \cap \partial B(a,r_a)} |Du|^p \, \mathcal{H}^{(n-1)} \right)^{(n-1)/p}$$
$$\leq c\, \varepsilon \, ||Du||^{n-1}_{L^p(\partial B(a,r_a))}$$

which proves (5.1.3).

We now consider a map $u \in \operatorname{cart}^1_+(\Omega, \hat{\mathbf{R}}^n) \cap W^{1,p}(\Omega.\hat{\mathbf{R}}^n)$, $p > n-1$, and we extend u on all of Ω as a set function in two ways by means of the following definition, compare [24].

Definition 5.3.1. *Let $u \in \mathrm{cart}^1_+(\Omega, \widehat{\mathbf{R}}^n)$ and let $x \in \Omega$. We define*

$$A_u(x) := \cap\{\overline{G} \mid G \text{ open}, \theta(u^{-1}(G), x) = 1\},$$
$$F_u(x) := \cap\{\overline{\overline{u}(\overline{B(x,r)} \cap \mathcal{A}_u)} \mid r > 0\},$$

where $\theta(A, x)$ denotes the Lebesgue density of A at x.

The next theorem collects a few pointwise properties in terms of the extensions $A_u(x)$ and $F_u(x)$.

Theorem 5.3.1. *Let $u \in \mathrm{cart}^1_+(\Omega, \widehat{\mathbf{R}}^n) \cap W^{1,p}(\Omega, \widehat{\mathbf{R}}^n)$, $p > n-1$. For $a \in \Omega$ denote by $r_a :=$ dist $(a, \partial\Omega)$ and by $R_a \subset (0, r_a)$ the set of $r \in (0, r_a)$ such that* (i), (ii), *and* (iii) *of Lemma* 5.3.1 *are satisfied and, moreover, the following additional conditions hold:*

$$\partial(G_u \llcorner B(a,r) \times \widehat{\mathbf{R}}^n) = G_{\tilde{u}_{\to|\partial B(a,r)}}, \tag{5.3.2}$$
$$\mathcal{H}^{n-1}(\partial B(a,r) \backslash \mathcal{R}_u) = 0, \tag{5.3.3}$$
$$\int_{\partial B(a,r) \cap \mathcal{R}_u} |D\tilde{u}|^p \, d\mathcal{H}^{n-1} < \infty. \tag{5.3.4}$$

Then we have:

(i) *$A_u(x)$ and $F_u(x)$ are closed sets, $A_u(x)$ is nonempty, and*

$$A_u(x) = \{\overline{u}(x)\} \qquad \text{if} \quad x \in \mathcal{A}_u.$$

(ii) $|(0, r_a) \backslash R_a| = 0.$

(iii) *Let $\Omega_1 \subset \Omega$ and $\mathcal{U} \subset \widehat{\mathbf{R}}^n$ be open sets. Then the set $\{x \in \Omega_1 \mid F_u(x) \subset \mathcal{U}\}$ is open. In particular, the function $x \to$ diam $F_u(x)$ is upper-semicontinuous in Ω.*

(iv) *Let $r \in R_a$ and let*

$$\widehat{\mathbf{R}}^n \backslash \overline{u}(\partial B(a,r)) = \bigcup_k W_k$$

be the decomposition into connected components. Then

$$\overline{\overline{u}(\overline{B(a,r)} \cap \mathcal{A}_u)} = \bigcup \{W_k \mid \overline{u}(B(a,r) \cap \mathcal{A}_u) \cap W_k \neq \emptyset\} \cup \overline{u}(\partial B(a,r)). \tag{5.3.5}$$

(v) *For all $a \in \Omega$, $F(a)$ is a connected and compact set.*

(vi) *For all* $r \in R_a$

$$\begin{aligned} \operatorname{diam} \overline{u}(\overline{B(a,r)} \cap \mathcal{A}_u) &= \operatorname{osc}(\overline{u}, \partial B(a,r)), \\ \operatorname{diam} F_u(a) &= \lim_{\substack{r\to 0 \\ r\in R_a}} \operatorname{osc}(\overline{u}, \partial B(a,r)). \end{aligned}$$

(vii) *If* diam $F_u(a) = 0$, *then* $a \in \mathcal{A}_u$ *and* $\overline{u}$ *is continuous at* a. *Conversely, if*

$$\lim_{\substack{x\to a \\ x\in\mathcal{A}_u}} \overline{u}(x) \ \textit{exists},$$

then diam $F_u(a) = 0$ *and* $a \in \mathcal{A}_u$.

(viii) *If* diam $F_u(a) > 0$, *then*

$$\int_0^{r_a} \left(r^{p-n} \int_{\partial B(a,r)\cap\mathcal{R}_u} |D\tilde{u}|^p \, d\mathcal{H}^{n-1} \right) dr = +\infty.$$

Proof. Part (ii) follows from Lemma 5.3.1 and Theorem 5.3.1.

Parts (i) and (iii) are consequences of the definitions of $A_u(x)$ and $F_u(x)$.

Part (iii). By the definition of R_a, u is everywhere defined and continuous on $\partial B(a,r)$ so that $\overline{u}(\partial B(a,r))$ is a compact set and we also have

$$\overline{u}(\partial B(a,r)) = \overline{\overline{u}(\partial B(a,r) \cap \mathcal{R}_u)} = \operatorname{spt} \partial (G_u \llcorner B(a,r) \times \widehat{\mathbf{R}}^n)$$

by (5.3.3) and (5.3.4). If $\overline{u}(B(a,r) \cap \mathcal{A}_u) \cap W_k \neq \emptyset$ we have $m_k = \deg(u, B(a,r), y) \geq 1$ for a.e. $y \in W_k$. By Theorem 5.3.1 (i) and (ii) we obtain

$$|W_k \backslash \overline{u}(B(a,r) \cap \mathcal{A}_u)| = 0.$$

Whence

$$\overline{\overline{u}(B(a,r) \cap \mathcal{A}_u)} \supset W_k.$$

This clearly shows one inclusion in (5.3.5), taking into account that $\overline{u}(\partial B(a,r))$ is closed, in fact compact. Observing now that there is only one unbounded component, say W_{k_0}, as $\overline{u}(\partial B(a,r))$ is compact, and that for all $k \neq k_0$ we have $\partial W_k \subset \overline{u}(\partial B(a,r))$, we easily infer the other inclusion. Notice that we have also proved that the set in (5.3.5) is compact.

(iv) It follows from the proof of (iii).

(v) Let $x \in W_k$ and $y \in W_\ell$, $k, \ell \neq k_0$. W_k and W_ℓ are bounded, therefore the straight line through x and y must intersect $\partial W_k \cup \partial W_\ell \subset \overline{u}(\partial B(a,r))$ on points which are outside the segment $[x, y]$ and opposite with respect to $[x, y]$. From this remark we readily infer the conclusions.

(vi) It immediately follows from (v).

(vii) If diam $F_u(a) = b > 0$, we infer

$$\operatorname{osc}(\overline{u}, \partial B(a,r)) \geq b \qquad \text{for a.e.} \quad r \in R_a$$

from (v).

The claim then follows from Lemma 5.3.1(iii), taking into account (ii). ■

A simple consequence of Theorem 5.3.1 is the following:

Theorem 5.3.2. (Vodopianov–Goldstein). *Let $u \in W^{1,n}(\Omega, \widehat{\mathbf{R}}^n)$ satisfy* $\det Du > 0$ *a.e. in Ω. Then $\mathcal{A}_u = \Omega$ and $\overline{u}$ is continuous in Ω.*

Proof. By a simple approximation argument, [5], [4], we obtain that $\partial G_u \llcorner \Omega \times \widehat{\mathbf{R}}^n = 0$, hence $u \in \operatorname{cart}^1_+(\Omega, \widehat{\mathbf{R}}^n)$. Theorem 5.3.1(vii) then implies diam $F_u(a) = 0$ for all $a \in \Omega$, and Theorem 5.3.1(iv) that every $a \in \Omega$ belongs to $\mathcal{A}_u$ and that $\overline{u}$ is continuous at every $a \in \Omega$. ■

Of course, we cannot expect that maps in $\operatorname{cart}^1_+(\Omega, \hat{\mathbf{R}}^n) \cap W^{1,p}(\Omega, \hat{\mathbf{R}}^n)$, $p > n-1$, be continuous everywhere. However we have, compare [24] and [18].

Theorem 5.3.3. *Let $u \in \operatorname{cart}^1_+(\Omega, \widehat{\mathbf{R}}^n) \cap W^{1,p}(\Omega, \widehat{\mathbf{R}}^n)$. Set*

$$\sum := \{x \in \Omega \mid \operatorname{diam} F_u(x) > 0\}.$$

Then we have

$$\dim_{\mathcal{H}} \sum \leq n - p, \qquad \Omega \setminus \sum \subset \mathcal{A}_u,$$

and $\overline{u}$ is continuous at each $x \in \Omega \setminus \sum$.

Proof. The argument used is taken from [18], see also [24]. Because of Theorem 5.3.2 we can assume $n - 1 < p < n$. We consider the L^1-function

$$w(x) := \begin{cases} |D\tilde{u}(x)|^p & \text{if } x \in \mathcal{R}_u, \\ 0 & \text{otherwise,} \end{cases}$$

and its Riesz potential

$$(K_{n-p} * w)(a) := \int_{\mathbf{R}^n} \frac{w(x)}{|a-x|^{n-p}}\, dx.$$

From Theorem 5.3.1(vii) we infer

$$(K_{n-p} * w)(a) = +\infty, \qquad \forall a \in \sum.$$

As by Theorem 5.3.1(vi) we can write $\sum = \cup \sum_k$, $\sum_k$ compact, a result of D. Frostman on Riesz's potentials, compare, e.g., [17, Theorems 5.9 and 5.13], yields $\mathcal{H}^{n-p+\delta}(\sum) = 0$ for all $\delta > 0$. ■

Example 5.3.1. Consider the map $\Omega \to \widehat{\mathbf{R}}^2$, defined in $\Omega\setminus\{(0,0)\}$, $\Omega = (-1,1)\times(-2,2) \subset \mathbf{R}^2$, by

$$u^1(x) := x_1,$$
$$u^2(x) := \begin{cases} x_2\,|x_1|^{-\alpha} & \text{if } |x_2| \le |x_1|^\alpha,\ |x_1| > 0,\\ \left(2 - \dfrac{2-|x_2|}{2-|x_1|^\alpha}\right)\operatorname{sgn} x_2 & \text{if } |x_1|^\alpha \le |x_2|,\ |x_2| > 0, \end{cases}$$

where $\alpha > 0$. It is easily seen that

$$u \in \operatorname{cart}^p_+(\Omega, \widehat{\mathbf{R}}^2) \cap W^{1,p}(\Omega, \widehat{\mathbf{R}}^2)$$

for all p with

$$p < 1 + \min\left(\alpha, \frac{1}{\alpha}\right).$$

Notice that u is a smooth diffeomorphism of $\Omega\setminus\{(0,0)\}$ onto $\Omega\setminus(\{0\}\times[-1,1])$.
We easily check that

$$F_u(0,0) = \{0\}\times[-1,1],$$

while

$$A_u(0,0) = \begin{cases} (0,0) & \text{if } \alpha < 1,\\ \{0\}\times[-1,1] & \text{if } \alpha = 1,\\ \{(0,1)\}\cup\{(0,1)\} & \text{if } \alpha > 1. \end{cases}$$

We then see that:

1. for $\alpha > 1$, we have $\mathcal{A}_u = \Omega$, $\overline{u}(0,0) = (0,0)$, but the map $\overline{u}$ is not continuous at $(0,0)$;
2. for $\alpha = 1$, we have $\mathcal{A}_u = \Omega\setminus\{(0,0)\}$ and
$$A_u(0,0) = F_u(0,0) = \{0\}\times[-1,1];$$
3. for $\alpha > 1$, $A_u(0,0)$ is not connected in contrast with $F_u(0,0)$ which is always connected.

Notice finally that $F_u(0,0)$ fills up the hole in $\Omega\setminus(\{0\}\times[-1,1])$ and that $F_u(0,0)\cap F_u(x) = \emptyset$ for all $x \ne (0,0)$.

5.4. The Range of Maps in $\operatorname{cart}^1_+(\Omega, \hat{\mathbf{R}}^n) \cap W^{1,p}(\Omega, \hat{\mathbf{R}}^n)$, $p > n-1$

It is the aim of this last section to prove that the range of locally weakly invertible maps with p-summable derivatives, $p > n-1$, is *a.e.-open*, and, in particular, that for every weak diffeomorphism $u \in \widetilde{\text{dif}}^{1,1}(\Omega, \hat{\mathbf{R}}^n) \cap W^{1,p}(\Omega, \hat{\mathbf{R}}^n)$, $p > n-1$, $\tilde{u}$ is an *a.e.-homeomorphism* from Ω into $\tilde{u}(\Omega)$.

Definition 5.4.1. We say that $A \subset \mathbf{R}^n$ is *a.e.-open* iff there exists an open set G such that $A = G$ a.e.

We will define *interior points*, and *accumulation points* of A by

$$\begin{aligned} \text{Int}\,(A) &:= \{x \in \mathbf{R}^n \mid \exists r : B(x,r) \subset A \text{ a.e.}\}, \\ \text{Der}\,(A) &:= \{x \in \mathbf{R}^n \mid |B(x,r) \cap A| > 0, \ \forall r > 0\}. \end{aligned}$$

We will also define

$$d_{\text{ess}}(x, A) := \sup\{\text{dist}(x, A \backslash N) \mid |N| = 0\}.$$

Though a.e.-open sets do not define a topology, the following two propositions show that they provide a reasonable structure, see [15] and [16].

Proposition 5.4.1. *We have*

(i) $\text{Int}(A)$ *is open and* $\text{Int}\,(A) \subset A$ a.e. *for any* $A \subset \mathbf{R}^n$. *If* G *is open and* $G \subset A$ a.e., *then* $G \subset \text{Int}(A)$; A *is* a.e.-*open iff* $A = \text{Int}(A)$ a.e. *Finally, for all* A

$$\text{Int}(A) = \cup\{G \mid G \text{ open}, G \subset A \text{ a.e.}\}.$$

(ii) *Finite intersections and arbitrary union of a.e.-open sets are a.e.-open*

(iii) *G is a.e.-open iff for a.e. $x \in G$ there exists $r > 0$ such that $B(x, r) \subset G$ a.e.*

(iv) $\text{Int}(A)$ *and* $\text{Der}(A)$ *depend only on the class* $[A]$ *of* A, *i.e.,*

$$\begin{aligned} \text{Int}(A_1) &= \text{Int}(A_2) \quad \text{if} \quad A_1 = A_2 \text{ a.e.}, \\ \text{Der}(A_1) &= \text{Der}(A_2) \quad \text{if} \quad A_1 = A_2 \text{ a.e.} \end{aligned}$$

Moreover

$$\text{Der}(A) = \{x \in \mathbf{R}^n \mid d_{\text{ess}}(x, A) = 0\}.$$

Proof. It is clear that $\text{Int}(A)$ is open for all $A \subset \mathbf{R}^n$. For a given A we set

$$\mathcal{B} := \{B(x,r) \mid x \in \text{Int}(A),\ B(x,r) \subset A \text{ a.e.}\}.$$

We may choose a countable subcovering $\mathcal{B}' \subset \mathcal{B}$ of $\text{Int}(A)$ such that $\cup\mathcal{B}' \subset A$ a.e. This proves that $\text{Int}(A) \subset A$ a.e. The other claims are trivial. ■

Definition 5.4.2. *Let $\Omega \subset \mathbf{R}^n$ be a measurable set and let $u : \Omega \to \widehat{\mathbf{R}^n}$ be a measurable function. We shall say that:*

u is a.e.-continuous iff $u^{-1}(G)$ is a.e.*-open for all* a.e.*-open sets G;*
u is a.e.-open iff $u(G)$ is a.e.*-open whenever G is* a.e.*-open;*
u is a.e.-continuous at $x \in \Omega$ iff for all $\varepsilon > 0$ there is $\delta > 0$ such that

$$u(B(x,\delta)) \subset B(u(x),\varepsilon) \quad \text{a.e. in } \hat{\mathbf{R}}^n;$$

u is an a.e-homeomorphism from Ω onto $\widehat{\Omega}$ iff u is a.e.*-continuous and* a.e.*-open, u is one to one a.e., i.e., there exists $\Omega_1 \subset \Omega$ such that*

$$|\{x \in \Omega_1 \mid \exists\, y \in \Omega_1,\ y \neq x,\ u(x) = u(y)\}| = 0$$

and $|\widehat{\Omega} \Delta u(\Omega_1)| = 0$.

Finally, we will say that u satisfies the double Lusin's condition iff

$$|u(A)| = 0 \quad \Longleftrightarrow \quad |A| = 0 \qquad \text{for} \quad A \subset \Omega.$$

Proposition 5.4.2. *Let Ω be a measurable set in $\mathbf{R}^n$ and let $u : \Omega \to \widehat{\mathbf{R}}^n$ be a measurable function satisfying the double Lusin condition. Then u is* a.e.*-continuous if and only if u is* a.e.*-continuous at* a.e. $x \in \Omega$.

Proof. Suppose u is a.e.-continuous in Ω. For $x \in \Omega$ we set

$$R(x) := \inf\{r > 0 \mid x \in \operatorname{Int}(u^{-1}(B(u(x),r)))\}.$$

If $R = 0$ a.e. in Ω, then u is a.e.-continuous at a.e. $x \in \Omega$. Assume on the contrary that

$$|\{x \in \Omega \mid R(x) > 0\}| > 0.$$

Then we find $\varepsilon > 0$ and $A_1 \subset \Omega_1$, $|A_1| > 0$, such that

$$R(x) \geq \varepsilon, \qquad \forall x \in A_1,$$

where we denoted by A_1 the set of points of density 1 of A.

Clearly the collection of balls

$$\mathcal{B} := \{B(u(x),r) \mid x \in A_1,\ r < \varepsilon/3\}$$

covers $u(A_1)$. We can then find a sequence $\{x_k\} \subset A_1$and $r_k \in (0, \varepsilon/3)$ such that

$$\bigcup_{k=1}^{\infty} B(u(x_k), r_k) \supset u(A_1)$$

therefore

$$\bigcup_k u^{-1}(B(u_k(x_k), r_k)) \supset A_1.$$

Then there is $k \in \mathbf{N}$ such that we have

$$|u^{-1}(\widehat{B}) \cap A_1| > 0, \qquad \widehat{B} := B(u(x_k), r_k).$$

By our assumption on u, $u^{-1}(\widehat{B})$ is a.e.-open, thus there exists an open set G with $G \subset u^{-1}(\widehat{B}) \cup N$ for some null set N; hence

$$|G \cap u^{-1}(\widehat{B}) \cap A_1| > 0.$$

Consequently, there exists $x_0 \in G \cap u^{-1}(\widehat{B}) \cap A_1$ such that $u(x_0) \in \widehat{B} = B(u(x_k), r_k)$ and therefore

$$x_0 \in \operatorname{Int}(u^{-1}(\widehat{B})) \subset \operatorname{Int}(u^{-1}(B(u(x_0), 2r_k)).$$

It follows, in particular, that $R(x_0) \leq 2r_k < \varepsilon$, which contradicts $x_0 \in A_1$.

This shows that u is a.e.-continuous at a.e. $x \in \Omega$. The converse follows easily from Proposition 5.4.1. ∎

We now prove;

Theorem 5.4.1. *Let Ω be a bounded open set in $\mathbf{R}^n$ and let $u \in \operatorname{cart}^1_+(\Omega, \widehat{\mathbf{R}}^n)$. Suppose moreover that $u \in W^{1,p}(\Omega, \widehat{\mathbf{R}}^n)$ for some $p > n-1$. Then $\tilde{u}$ is a.e.-open.*

Proof. Since $\tilde{u}$ has the Lusin property (N), it suffices to prove that $\tilde{u}(G)$ is a.e.-open for each open set $G \subset \Omega$. Fix $a \in G$ and $r > 0$ such that $B(a,r) \subset G$. By Theorem 5.4.1, we can select r_a, $0 < r_a < r$, in such a way that

$$\partial(G_u \llcorner B(a, r_a) \times \hat{\mathbf{R}}^n) = G_{\tilde{u}, \partial B(a, r_a)},$$

in particular,

$$\operatorname{spt} \widehat{\pi}_\# \, \partial(G_u \llcorner B(a,r) \times \widehat{\mathbf{R}}^n) \subset \overline{\tilde{u}(\partial B(a,r) \cap \mathcal{R}_u)};$$

moreover, we can suppose that the trace $\overline{u}$ of u on $\partial B(a, r_a)$ belongs to $W^{1,p}(\partial B(a, r_a), \hat{\mathbf{R}}^n)$ and $\tilde{u} = \overline{u}$ $\mathcal{H}^{n-1}$-a.e. on $\partial B(a, r_a)$, and also that $\overline{u} : \partial B(a, r_a) \to \hat{\mathbf{R}}^n$ is continuous, and

$$\mathcal{H}^{n-1}(A) = 0 \qquad \text{provided} \qquad A \subset \partial B(a, r_a), \qquad \mathcal{H}^{n-1}(A) = 0. \tag{5.4.1}$$

Denote now by

$$F := \overline{u}(\partial B(a, r_a)).$$

F is compact, and taking into account that $\tilde{u}(\partial B(a, r_a))$ is $\mathcal{H}^{n-1}$-rectifiable and (5.4.1)

$$|F| = 0. \tag{5.4.2}$$

We now decompose the complement of F into connected components

$$\widehat{\mathbf{R}}^n \backslash F = \bigcup_k W_k$$

and denote by $\bigcup' W_k$ the union of the W_k where the degree of u on $B(a, r_a)$ is a.e. a constant greater than 1. Since $\det Du > 0$ a.e. we conclude that

$$\begin{aligned} \tilde{u}(B(a, r_a)) \backslash F &= \{ y \in W_k \mid \deg(u, B(a, r_a), y) \geq 1 \} \\ &= \bigcup{}' W_k \text{ a.e.} \end{aligned}$$

and the conclusion follows from (5.4.2). ■

Writing $G_u = \tau(\mathcal{M}, 1, \vec{G}_u)$, since the projection of $\mathcal{M}$ on $\hat{\mathbf{R}}^n$, $\operatorname{im} G_u$, has positive measure and the a.e. inverse of u, $\hat{u}$, is almost everywhere approximately differentiable on $\operatorname{im} G_u$, compare [10], Theorem 5.4.1 then yields at once:

Corollary 5.4.1. *Let Ω be a bounded set in $\mathbf{R}^n$ and let $u \in \operatorname{dif}^{1,1}(\Omega, \widehat{\mathbf{R}}^n) \cap W^{1,p}(\Omega, \widehat{\mathbf{R}}^n)$ for some $p > n - 1$. Then $\tilde{u}$ is an a.e.-homeomorphism from Ω onto $\tilde{u}(\Omega)$.*

References

[1] Ball, J.M. and Murat, F. $W^{1,p}$-quasiconvexity and variational problems for multiple integrals. *J. Funct. Anal.*, (1988), 225–253.

[2] Brezis, H., Coron, J. and Lieb, E. Harmonic maps with defects. *Comm. Math. Phys.*, **107** (1986), 649–705.

[3] Giaquinta, M., Modica, G. and Souček, J. *Cartesian Currents in the Calculus of Variations.* Springer-Verlag, Heidelberg, to appear.

[4] Giaquinta, M., Modica, G., and Souček, J. Cartesian currents and variational problems for mappings into spheres. *Ann. Scuola Norm. Sup. Pisa*, **16** (1989), 393–485.

[5] Giaquinta, M., Modica, G., and Souček, J. Cartesian currents, weak diffeomorphisms and existence theorems in nonlinear elasticity. *Arch. Rational Mech. Anal.*, **106** (1989), 97–159. Erratum and addendum. *Arch. Rational Mech. Anal.*, **109** (1990), 385–392.

[6] Giaquinta, M., Modica, G., and Souček, J. The Dirichlet energy of mappings with values into the sphere. *Manuscripta Math.*, **65** (1989), 489–507.

[7] Giaquinta, M., Modica, G., and Souček, J. Liquid crystals: Relaxed energies, dipoles, singular lines and singular points. *Ann. Scuola Norm. Sup. Pisa*, **17** (1990), 415–437.

[8] Giaquinta, M., Modica, G., and Souček, J. The Dirichlet integral for mappings between manifolds: Cartesian currents and homology. *Math. Ann.*, **294** (1992), 325–386.

[9] Giaquinta, M., Modica, G., and Souček, J. Variational problems for the conformally invariant integral $\int |du|^n$. In (C. Bandle, J. Bemelmans, C.M. Grüter, and J.S.J. Paulin, eds.). *Progress in Partial Differential Equations: Calculus of Variations, Applications.* Pitman Research Notes in Mathematics, vol. **267**, pp. 27–47. Pitman, London, 1992.

[10] Giaquinta, M., Modica, G., and Souček, J. A weak approach to finite elasticity. (1993). To appear. *Calc. Var. Partial Differential Equations* **2** (1994), 65–100.

[11] Giaquinta, M., Modica, G., and Souček, J. Connectivity properties of the range of weak diffeomorphisms. Preprint, 1993. *Ann. Inst. H. Poincaré Anal. Nonlinear*, **12** (1995), 61–73.

[12] Giaquinta, M., Modica, G., and Souček, J. Graphs of finite mass which cannot be approximated in area by smooth graphs, 1993. *Manuscripta Math.*, **78**, 259–271.

[13] Giaquinta, M., Modica, G., and Souček, J. Remarks on the degree theory, 1993. *J. Funct. Anal.*, **125** (1994), 172–200.

[14] Giusti, E. Precisazione delle funzioni $H^{1,p}$ e singolarità delle soluzioni deboli di sistemi ellittici non lineari. *Boll. Un. Mat. Ital.*, **2** (1969), 71–76.

[15] Goffman, C., Neugebauer, J., and Nishiura, C.T. Density topology and approximate continuity. *Duke Math. J.*, **28** (1961), 497–505.

[16] Goffman, C. and Waterman, D. Approximately continuous transformations. *Proc. Amer. Math. Soc.*, **12**, (1961), 116–121.

[17] Hayman, W. and Kennedy, P. *Subharmonic Functions*. Academic Press, New York, 1976.

[18] Malý, J. Private communication.

[19] Malý, J. L^p-approximation of Jacobians. *Comment. Math. Univ. Carolin.*, **32** (1991), 659–666.

[20] Müller, S. Weak continuity of determinants and nonlinear elasticity. *C. R. Acad. Sci. Paris*, **307** (1988), 501–506.

[21] Müller, S., Tang, Q., and Yan, B. On a new class of elastic deformations not allowing cavitation. *Ann. Inst. H. Poincaré Anal. Nonlinear*, **11** (1994), 217–243.

[22] Reshetnyak, Y. *Space Mappings with Bounded Distorsion*. Transl. Math. Monographs. American Mathematical Society, Providence, RI, 1989.

[23] Tang, Q. Almost everywhere injectivity in nonlinear elasticity. *Proc. Roy. Soc. Edinburgh*, **109A** (1988), 79–95.

[24.] Šverák, V. Regularity properties of deformations with finite energy. *Arch. Rational Mech. Anal.*, **100** (1988), 105–127.

[25] Ziemer, W.P. *Weakly Differentiable Functions*. Springer-Verlag, New York, 1989.

6
Some Recent Results on Saint-Venant's Principle

R.J. Knops and C. Lupoli

6.1. Introduction

Saint-Venant's principle has been accepted by structural engineers, numerical analysts, and others as a reliable guide to how far edge and boundary effects penetrate into a body. Nevertheless, this conjecture by Saint-Venant in 1855, even though based upon plausible intuitive argument, has from the beginning provoked scepticism and the history of the subject has been characterized by attempts to formulate a precise mathematical definition supported by rigorous mathematical proof. Two main approaches may be discerned. One group investigates relevant properties of exact solutions: within this category are authors such as Boussinesq [4], Clebsch [5], Dougall [7], Synge [41], von Mises [32], Sternberg [39], and by slight extension, Mielke [30], [31] who has applied center manifold arguments. The other main area of study considers energy, and early contributors include Zanaboni [43], Goodier [23], [18], and Dou [6], the classic paper being by Toupin [42]. Later writers are, for example, Fichera [9], [10], [11], Oleinik et al. [24]–[28], [36], Payne and Horgan and their respective coworkers. The subject has been comprehensively surveyed by Gurtin [19], Maisonneuve [29], and Horgan and Knowles [20].

Saint-Venant's principle was originally formulated within the context of linear elasticity and hence properly belongs to a study of elliptic partial differential equations. Yet the literature of this field rarely, if ever, refers to the principle despite its undoubted practical significance. It is worth briefly seeking an explanation for this apparent neglect. Let us consider the simple example of Laplace's equation on a semi-infinite three-dimensional strip. The solution contains both exponentially increasing and decreasing terms which, we note, provide an example of the Phragmén–Lindelöf principle. The imposition of a boundedness condition upon the solution eradicates the growth element and there remains only the exponentially decreasing terms consistent with Saint-Venant's principle. In this sense, therefore, Saint-Venant's principle should be regarded as part of a Phragmén–Lindelöf principle which has, of course, been extensively studied.

Many of the previous investigations of Saint-Venant's principle have apparently failed to appreciate this relationship. Indeed treatments involving volume energy measures, in requiring such measures to be bounded, exclude from the start any possibility of growth behavior and thus of a link being established with a Phragmén–Lindelöf principle.

Recently, however, a series of papers [12]–[16] and [17], has developed an approach used by Payne and Weinberger [37] that employs a variant of the energy argument and permits the discussion of both growth and decay behavior and hence of demonstrating the relationship between both principles. The technique uses a surface energy flux instead of a volume energy and derives a first-order differential inequality for the flux. Integration then yields the desired behavior. The method is briefly described in the next section, and can be applied to a wide variety of problems, not necessarily of elliptic type. We here confine attention to three applications, namely those of linearized nonhomogeneous elasticity (Section 6.3), linearized nonhomogeneous thermoelasticity (Section 6.4), and linearized nonhomogeneous viscoelasticity (Section 6.5), and for each sketch the derivation of the basic differential inequality which is more fully discussed in Section 6.2. For the sake of simplicity, we treat only the three-dimensional semi-infinite cylinder with the lateral surface held rigidly fixed, but extensions are possible to the cylinder with free lateral boundary, and to other geometries including sectorial and exterior regions, and the whole and half-space.

Throughout, existence of a smooth solution is assumed, and the conventions are adopted of summation over repeated subscripts, and the comma notation to indicate partial differentiation. Latin suffixes take values in the range $1, 2, 3$ while Greek suffixes range over $1, 2$.

This paper is in part extracted from the thesis by C. Lupoli and the support of the SERC is gratefully acknowledged.

6.2. Notation and Basic Results

We consider a three-dimensional prismatic cylinder Ω with plane base, and select a rectangular system of Cartesian coordinates such that the base of the cylinder lies in the (x_1, x_2)-coordinate plane and contains the origin, while the generators of the cylinder are parallel to the positive x_3-axis. The cylinder is semi-infinite in length. We suppose that $D(x_3)$ represents the bounded cross-section of the cylinder at distance x_3 from the plane base, and that the boundary $\partial D(x_3)$ of each $D(x_3)$ is sufficiently smooth to admit application of the divergence theorem in the plane of the cross-section.

The cylinder is occupied by a material which is in equilibrium subject to specific displacements over the base. There is no body-force, nor is the asymptotic behavior of the displacement, u_i, prescribed for large values of the axial variable x_3. The stress, σ_{ij}, in the material therefore satisfies the equilibrium equations

$$\sigma_{ij,j} = 0, \qquad x \in \Omega, \tag{6.2.1}$$

while the displacement satisfies the boundary condition

$$u_i = 0, \qquad x \in \partial D(x_3) \times [0, \infty]. \tag{6.2.2}$$

To these equations must be appended the stress constitutive relation describing the composition of the particular material under consideration. Three examples are presented in the succeeding sections. Even so, the essence of the present technique can be described using only equations (6.2.1) and (6.2.2). Thus, let us introduce the energy flux defined by

$$H(x_3) = \int_{D(x_3)} \sigma_{i3} u_i \, dx_1 \, dx_2. \tag{6.2.3}$$

Then we have for any constant positive h,

$$H(x_3 + h) - H(x_3) = \int_{x_3}^{x_3+h} \int_{D(\eta)} \sigma_{ij} u_{ij} \, dx_1 \, dx_2 \, d\eta, \tag{6.2.4}$$

where (6.2.1), (6.2.2) have been used together with an integration by parts. It immediately follows that

$$H'(x_3) = \int_{D(x_3)} \sigma_{ij} u_{ij} \, dx_1 \, dx_2. \tag{6.2.5}$$

Now let us suppose that a positive function $P(x_3)$ can be found such that for positive constants k_1, k_2

$$\text{(i)} \qquad H'(x_3) \geq k_1 P(x_3), \tag{6.2.6}$$

$$\text{(ii)} \qquad |H(x_3)| \leq k_2 P(x_3). \tag{6.2.7}$$

On combining these inequalities, we may conclude that

$$|H(x_3)| \leq k^{-1} H'(x_3), \qquad k = k_1/k_2, \tag{6.2.8}$$

which is the required fundamental inequality. Its treatment, the same for all cases, is outlined here.

First, let us suppose that $H(0) > 0$. Then, by (6.2.6), it follows that $H(x_3) > 0$ for $x_3 \geq 0$, and hence (6.2.8) yields

$$H'(x_3) - kH(x_3) \geq 0, \tag{6.2.9}$$

which immediately may be integrated to give

$$H(x_3) \geq H(0) \exp(kx_3). \tag{6.2.10}$$

Thus, $H(x_3)$ becomes exponentially unbounded as $x_3 \to \infty$; and hence by (6.2.8) and (6.2.5) we conclude that

$$\lim_{x_3 \to \infty} \int_{D(x_3)} \sigma_{ij} u_{ij} \, dx_1 \, dx_2 \tag{6.2.11}$$

is likewise unbounded. In the examples selected here, it will be evident that $H'(x_3)$ is related to the energy and hence these conclusions may alternatively be expressed in terms of this quantity. The validity of these results does not depend upon $H(0) > 0$. It suffices that $H(\xi) > 0$ for some $\xi \geq 0$. The last remark implies that when the limit (6.2.11) is bounded, then necessarily $H(x_3) \leq 0$ for $x_3 \geq 0$. The relevant component of inequality (6.2.8) then gives

$$H'(x_3) + kH(x_3) \geq 0, \tag{6.2.12}$$

which on integration leads to

$$-H(x_3) \leq -H(0)\exp(-kx_3) \tag{6.2.13}$$

and thus

$$\lim_{x_3\to\infty} H(x_3) = 0. \tag{6.2.14}$$

On defining the energy to be

$$E(x_3) = \int_{x_3}^{\infty}\int_{D(\eta)} \sigma_{ij}u_{ij}\, dx_1\, dx_2\, d\eta, \tag{6.2.15}$$

we may conclude from (6.2.4), (6.2.13), and (6.2.14) that

$$E(x_3) \leq E(0)\exp\{-kx_3\}. \tag{6.2.16}$$

In certain circumstances, the last inequality leads to an estimate for the cross-section mean square measure of the displacement. For example, let us suppose that

$$\lim_{x_3\to\infty}\int_{D(x_3)} u_iu_i\, dx_1\, dx_2 = 0. \tag{6.2.17}$$

Then

$$\begin{aligned}\int_{D(x_3)} u_iu_i\, dx_1\, dx_2 &= -2\int_{x_3}^{\infty}\int_{D(\eta)} u_iu_{i,3}\, dx_1\, dx_2\, d\eta\\ &\leq 2\lambda^{-1/2}\left[\int_{x_3}^{\infty}\int_{D(\eta)} u_{i,\alpha}u_{i,\alpha}\, dx_1\, dx_2\, d\eta\right.\\ &\quad\times\left.\int_{x_3}^{\infty}\int_{D(\eta)} u_{i,3}u_{i,3}\, dx_1\, dx_2\, d\eta\right]^{1/2},\\ &\leq \lambda^{-1/2}\int_{x_3}^{\infty}\int_{D(\eta)} u_{i,j}u_{i,j}\, dx_1\, dx_2\, d\eta,\end{aligned} \tag{6.2.18}$$

where we have appealed to the Schwarz and arithmetic geometric mean inequalities and the Poincaré inequality stated below. On further assuming the existence of a positive constant k_3 such that

$$\int_{D(x_3)} u_{i,j}u_{i,j}\,dx_1\,dx_2 \leq k_3 P(x_3), \tag{6.2.19}$$

we may conclude from (6.2.18), (6.2.19), (6.2.6), and (6.2.16) that

$$\int_{D(x_3)} u_i u_i\,dx_1\,dx_2 \leq \left(k_3/k_1\lambda^{1/2}\right) E(0)\exp\{-kx_3\}. \tag{6.2.20}$$

It is worth observing that inequality (6.2.8) holds irrespective of the cylinder's length and thus in particular remains valid for a cylinder of finite length when results similar to those established by Fichera [9], [10] may be recovered. Likewise, inequality (6.2.8) holds on the whole cylinder for which $x_3 \in (-\infty, \infty)$ and then it may be shown that only the trivial solution exists in the class of solutions possessing bounded energy.

Finally, for convenience, we state the *Poincaré inequality.*

Let λ denote the first eigenvalue in the two-dimensional clamped membrane problem for the uniform cross-section D, and let $\varphi(x_1, x_2)$ be a continuously differentiable function vanishing on ∂D. Then the following Poincaré inequality holds:

$$\lambda \int_D \varphi^2\,dx_1\,dx_2 \leq \int_D \varphi_{,\alpha}\varphi_{,\alpha}\,dx_1\,dx_2. \tag{6.2.21}$$

In the next three sections we demonstrate how the theory just outlined applies to examples drawn from linear continuum mechanics. The body is always assumed to be in equilibrium under zero body force and we suppose that equations (6.2.1), (6.2.2) are satisfied. The task is therefore to establish inequalities (6.2.6), (6.2.7) for the respective stress constitutive relation.

6.3. Linearized Elasticity

The relevant constitutive relation for the stress is

$$\sigma_{ij} = c_{ijkl}u_{k,l}, \tag{6.3.1}$$

where $c_{ijkl}(x)$ forms the elasticity tensor which is assumed to possess the major symmetry

$$c_{ijkl} = c_{klij}; \tag{6.3.2}$$

to be positive-definite in the sense that

$$c_0\psi_{ij}\psi_{ij} \leq c_{ijkl}\psi_{ij}\psi_{kl} \tag{6.3.3}$$

for positive uniform constant c_0 and second-order tensor ψ_{ij}, and to be bounded:

$$c_{ijkl}\xi_i\eta_j\xi_k\eta_l \leq c_1\xi_i\xi_i\eta_j\eta_j, \tag{6.3.4}$$

where c_1 is a positive uniform constant. Thus, we are dealing with the linearized theory of elasticity and *not* with classical linear elasticity for which the elasticities additionally enjoy the minor symmetries:

$$c_{ijkl} = c_{jikl} = c_{ijlk}.$$

Now, from (6.3.1) it follows immediately that

$$H'(x_3) = \int_{D(x_3)} c_{ijkl} u_{i,j} u_{k,l} \, dx_1 \, dx_2, \tag{6.3.5}$$

which under assumption (6.3.3) is positive-definite. Hence, in this case, $P(x_3) = H'(x_3)$, and we have proved (6.2.6). We establish (6.2.7) by noting that

$$|H(x_3)| = \left| \int_{D(x_3)} c_{ijkl} u_{k,l} n_j u_i \, dx_1 \, dx_2 \right|, \tag{6.3.6}$$

where n_i is the unit outward normal on $D(x_3)$. By virtue of (6.3.3), we may then apply Schwarz's inequality followed by assumption (6.3.4), Poincaré's inequality (6.2.21) and (6.3.3) to successively obtain

$$\begin{aligned} |H(x_3)| &\le \left[\int_{D(x_3)} c_{ijkl} u_{i,j} u_{k,l} \, dx_1 \, dx_2 \int_{D(x_3)} c_{ijkl} u_i n_j u_k n_l \, dx_1 \, dx_2 \right]^{1/2} \\ &\le (c_1/\lambda c_0)^{1/2} \int_{D(x_3)} c_{ijkl} u_{i,j} u_{k,l} \, dx_1 \, dx_2. \end{aligned} \tag{6.3.7}$$

Thus, on combining with (6.3.5) we have (6.2.8) with $k^{-1} = (c_1/\lambda c_0)^{1/2}$.

The energy (6.2.15) now becomes

$$E(x_3) = \int_{x_3}^{\infty} \int_{D(\eta)} c_{ijkl} u_{i,j} u_{k,l} \, dx_1 \, dx_2$$

and hence for solutions in the class of bounded energy, inequality (6.2.16) holds. Thus, on appeal to (6.3.3) and the Poincaré inequality, we may conclude that (6.2.17) is valid and hence we are led to the decay estimate (6.2.20).

6.4. Linearized Thermoelasticity

We introduce the temperature T (> 0) which we assume to satisfy the equation

$$(a_{ij} T_{,i})_{,j} = 0, \qquad x \in \Omega, \tag{6.4.1}$$

where the coefficients $a_{ij}(x_3)$ are subject to the conditions

$$a_0 \xi_i \xi_i \le a_{ij} \xi_i \xi_j \le a_1 \xi_i \xi_i, \tag{6.4.2}$$

for positive uniform constants a_0, a_1. We also suppose that the temperature satisfies the boundary conditions

$$T = 0, \qquad x \in \partial D \times (0, \infty), \tag{6.4.3}$$

but as before do not specify the asymptotic behavior of T for large x_3.

The stress constitutive relation is given by

$$\sigma_{ij} = c_{ijkl} u_{k,l} + \kappa_{ij} T, \tag{6.4.4}$$

where the elasticities $c_{ijkl}(x)$ are assumed subject to conditions (6.3.2)–(6.3.4), and $\kappa_{ij}(x)$ is the conductivity tensor which is assumed to satisfy the boundedness condition:

$$M^2 \equiv \max_{x_3 \in [0,\infty]} \kappa_{ij}\kappa_{ij} < \infty. \tag{6.4.5}$$

Earlier studies of Saint-Venant's principle in thermoelasticity include those by Boley [1], [2], [3], Horgan et al. [21], [22], and Oleinik and Yosifian [35].

In order to establish an inequality of the form (6.2.8) we must modify the function $H(x_3)$ defined by (6.2.3) by the addition of a further term. Thus, we consider

$$W(x_3) = H(x_3) + \gamma \int_{D(x_3)} a_{i3} T_{,i}\, T \, dx_1\, dx_2, \tag{6.4.6}$$

where γ is a positive constant to be optimally determined. We proceed to derive inequalities (6.2.6), (6.2.7) for $W(x_3)$ and hence inequality (6.2.8).

By means of Schwarz's inequality, Poincaré's inequality, and the arithmetic–geometric mean inequality it may be proved that

$$|W(x_3)| \leq K \int_{D(x_3)} (c_{ijkl} u_{i,j} u_{k,l} + a_{ij} T_{,i}\, T_{,j})\, dx_1\, dx_2, \tag{6.4.7}$$

where

$$K = \left(\frac{c_1}{\lambda c_0}\right)^{1/2} + \frac{[\sqrt{(A^2+1)} - A]M}{2\lambda (a_0 c_0)^{1/2}}, \tag{6.4.8}$$

$$A = \frac{(\lambda a_0 c_0)^{1/2}}{M} \left[\left(\frac{c_1}{c_0}\right) - \gamma \left(\frac{a_1}{a_0}\right)^{1/2}\right]. \tag{6.4.9}$$

and thus (6.2.7) is established.

Again, it easily follows that

$$W'(x_3) = \int_{D(x_3)} \left(c_{ijkl} u_{i,j} u_{k,l} + \kappa_{ij} T u_{i,j} + \gamma a_{ij} T_{,i}\, T_{,j}\right) dx_1\, dx_2, \tag{6.4.10}$$

from which for positive constant $\beta < 1$ we obtain

$$W'(x_3) \geq (1-\beta) \int_{D(x_3)} c_{ijkl} u_{i,j} u_{k,l}\, dx_1\, dx_2 + \left(\gamma - \frac{M^2}{4\lambda a_0 c_0 \beta}\right) \int_{D(x_3} a_{ij} T_{,i}\, T_{,j}\, dx_1\, dx_2. \tag{6.4.11}$$

Several choices of β and γ are possible, but we restrict attention to the case when the physical constants satisfy

$$M^2 < 4\lambda a_0 c_0 \tag{6.4.12}$$

and set

$$\gamma = 1, \qquad \beta = M/2(\lambda a_0 c_0)^{1/2} \tag{6.4.13}$$

so that

$$W'(x_3) \geq (1-\beta)\int_{D(x_3)} (c_{ijkl}u_{i,j}u_{k,l} + a_{ij}T_{,i}T_{,j})\,dx_1\,dx_2. \tag{6.4.14}$$

We remark that when $\gamma = 1$, the sign of A, given by (6.4.9), may be either positive or negative, but neither sign affects the validity of K in (6.4.8). Hence, on combining (6.4.14) and (6.4.7) we obtain inequality (6.2.8) with

$$k = K/1-\beta, \tag{6.4.15}$$

where β is given by (6.4.13) and (6.4.12) holds. When (6.4.12) is not satisfied, then inequality (6.2.8) may still be established but clearly for different choices of γ and β.

The discussion of inequality (6.2.8) follows that outlined in Section 6.2, except that when $W(\xi) > 0$, for some $\xi \geq 0$, it is only possible to prove that the *sum* of the mechanical and thermal energies becomes exponentially unbounded for large x_3, and not that either energy separately possesses this growth property.

6.5. Linearized Viscoelasticity

We consider the version of linearized viscoelasticity in which the stress constitutive relations are given by

$$\sigma_{ij}(x,t) = G_{ijkl}(0)u_{k,l}(x,t) + \int_0^t \frac{\partial G_{ijkl}(\eta)}{\partial \eta} u_{k,l}(x,t-\eta)\,d\eta, \tag{6.5.1}$$

where the tensor $G_{ijkl}(t)$ is nonincreasing in the sense that

$$\frac{\partial G_{ijkl}}{\partial t}(t)\psi_{ij}\psi_{kl} \leq 0, \qquad t \in (0,\infty), \tag{6.5.2}$$

for all ψ_{ij}. We also suppose that

$$G_{ijkl}(t) = G_{klij}(t), \qquad t \in (0,\infty), \tag{6.5.3}$$

and that for positive constants g_0, g_1 there holds

$$g_0\psi_{ij}\psi_{ij} \leq G_{ijkl}(t)\psi_{ij}\psi_{kl} \leq g_1\psi_{ij}\psi_{ij}, \qquad t \in (0,\infty), \tag{6.5.4}$$

for all ψ_{ij}. For our final assumption, we put

$$I_{ijkl}(t) = G_{ijkl}(0) - G_{ijkl}(t) \tag{6.5.5}$$

and suppose that for positive constant $\overline{g}$ we have

$$I_{ijkl}\psi_{ij}\psi_{kl} \geq \overline{g}\psi_{ij}\psi_{ij}, \qquad t \in (0, \infty), \tag{6.5.6}$$

for all ψ_{ij}.

We confine attention to *quasi-static deformations*, for which the stress satisfies the equilibrium equations (6.2.1), and we derive inequality (6.2.8) in terms of the function

$$V(x_3, t) = \int_0^t \int_{D(x_3)} \sigma_{i3}(x, s)u_i(x, s)\, dx_1\, dx_2\, ds. \tag{6.5.7}$$

By means of a procedure similar to that of Sections 6.3 and 6.4, it may be shown that the function V satisfies

$$|V(x_3, t)| \leq J \int_0^t \int_{D(x_3)} u_{i,j}(x, s)u_{i,j}(x, s)\, dx_1\, dx_2\, ds, \tag{6.5.8}$$

where

$$J = \left(\frac{g_1 - g_0}{\lambda}\right)^{1/2} \left[\frac{g_1}{g_0^{1/2}} + \frac{g_1 - g_0}{\overline{g}}\right]. \tag{6.5.9}$$

It may also be shown that

$$V'(x_3, t) \geq g_0 \int_0^t \int_{D(x_3)} u_{i,j}(x, s)u_{i,j}(x, s)\, dx_1\, dx_2\, ds, \tag{6.5.10}$$

so that we have established inequalities corresponding to (6.2.6) and (6.2 .7) and the basic inequality now follows with

$$k = g_0/J. \tag{6.5.11}$$

The analysis may be completed along lines indicated in Section 6.2.

Saint-Venant's principle for viscoelasticity has been previously discussed by Neapolitan and Edelstein [33], [34], and by Sternberg and Al Khozaie [40].

References

[1] Boley, B.A. The determination of temperature, stresses and deflections in two-dimensional thermoelastic problems. *J. Aerospace Sci.*, **23** (1956), 67–75.

[2] Boley, B.A. Some observations on Saint-Venant's principle. *Proc. 3rd Nat. Congr. Appl. Mech.*, ASME, (1958), pp. 259–264.

[3] Boley, B.A. Upper bounds and Saint-Venant's principle in transient heat conduction. *Quart. Appl. Mech.*, **18** (1960), 205–207.

[4] Boussinesq, J. *Application des Potentials à l'Étude de l'Equilibre et des Mouvements des Solides Élastiques.* Gauthier-Villars, Paris, 1885.
[5] Clebsch, A. *Théorie de l'Élasticité des Corps Solides.* (Translation of *Theorie der Elastizität fester Körper* by B de Saint-Venant and Flamant.) Dunod, Paris, 1883. Johnson Reprint Cooperation, New York, 1966.
[6] Dou, A. Upper estimate of the potential elastic energy of a cylinder. *Comm. Pure Appl. Maths.*, **19** (1966), 83–93.
[7] Dougall, J. An analytical theory of the equilibrium of an isotropic elastic rod of circular section. *Trans. Roy. Soc. Edinburgh*, **49** (1914), 895–978.
[8] Edelstein, W.S. A spatial decay estimate for the heat equation. *Z. Angew. Math. Phys.*, **20** (1969), 900–905.
[9] Fichera, G. Il principio di Saint-Venant: Intuizione dell'ingegnere e rigore del matematico. *Rend. Mat.*, **10** (Ser. VI) (1977), 1–24.
[10] Fichera, G. Remarks on Saint-Venant's principle. In *Complex Analysis and Its Applications.* I.N. Vekua Anniversary Volume, pp. 543–557, Nauka, Moscow, 1978.
[11] Fichera, G. Sull'esistenza e sul calcolo della soluzioni dei problemi al contorno relativi all'equilibrio di un corpo elastico. *Ann. Scuolo Norm. Sup. Pisa*, **4** (1950), 35–99.
[12] Flavin, J.N. and Knops, R.J. Some spatial decay estimates in continuum mechanics. *J. Elasticity*, **17** (1987), 249–264.
[13] Flavin, J.N. and Knops, R.J. Some decay and other estimates in two-dimensional linear elastostatics. *Quart. J. Mech. Appl. Math.*, **41** (1988), 223–238.
[14] Flavin J.N. and Knops, R.J. Some convexity considerations for a two-dimensional traction problem. *ZAMP*, **39** (1988), 166–176.
[15] Flavin, J.N., Knops, R.J., and Payne, L.E. Decay estimates for the constrained elastic cylinder of variable cross section. *Quart. Appl. Math.*, **47** (1989), 325–350.
[16] Flavin, J.N., Knops, R.J., and Payne, L.E. Asymptotic behavior of solutions to semi-linear elliptic equations on the half-cylinder. *Z. Angew. Math. Phys.*, **43** (1992), 405–421.
[17] Goodier, J.N. A general proof of Saint-Venant's principle. *Phil. Mag.*, (7), (1937), 607.
[18] Goodier, J.N. Supplementary note on "A general proof of Saint-Venant's principle." *Phil. Mag.*, (7), (1937), 24, 325.
[19] Gurtin, M.E. *The Linear Theory of Elasticity.* Handbuch der Physik, Vol. VI a/2, pp. 1–295. Springer-Verlag, New York, 1974.
[20] Horgan, C.O. and Knowles, J.K. Recent developments concerning Saint-Venant's principle. In *Advances in Applied Mechanics* (J.W. Hutchinson, ed.). vol. **23**, pp. 179–269. Academic Press, New York, 1983. See also: C.O. Horgan. Recent developments concerning Saint-Venant's principle: An update. *Appl. Mech. Rev.*, **42** (11), (1989), 295–302.
[21] Horgan, C.O., Payne, L.E., and Wheeler, L.T. Spatial decay estimates in transient heat conduction. *Quart. Appl. Math.*, **42** (1) (1984), 119–127.
[22] Horgan, C.O. and Wheeler, L.T. Spatial decay estimates for the heat equation via the maximum principle. *Z. Angew. Math. Phys.*, **27** (1976), 371–376.
[23] Knops, R.J., Rionero, S., and Payne, L.E. Saint-Venant's principle on unbounded regions. *Proc. Roy. Soc. Edinburgh*, **115A** (1990), 319–336.
[24] Kondratiev, V.A., Kopachek, I., and Oleinik, O.A. On the behaviour of weak solutions of second-order elliptic equations and the elasticity system in a neighbourhood of a boundary point. *Trudy Sem. Petrovsky.* **8** (1982), 135–152.
[25] Kondratiev, V.A. and Oleinik, O.A. On the asymptotics at infinity of solutions of elliptic systems with constant coefficients. *Uspekhi Mat. Nauk*, **40** (5), (1985), 233.

[26] Kondratiev, V.A. and Oleinik, O.A. On the asymptotic behaviour of solutions of systems of differential equations. *Uspekhi Mat. Nauk.*, **40** (5), (1985), 306.

[27] Kondratiev, V.A. and Oleinik, O.A. On the behaviour at infinity of solutions of elliptic systems with finite energy integral. *Arch. Rational Mech. Anal.*, **99** (1987), 75–89.

[28] Kondratiev, V.A. and Oleinik, O.A. Boundary value problems for the system of elasticity theory in unbounded domains. Korn's inequalities. *Uspekhi Mat. Nauk.*, **43** (1988), 55–98. English translation in *Russian Math. Surveys*, **43** (1988), 65–119.

[29] Maisonneuve, O. Sur le Principe de Saint-Venant. Thèse d'Etat, Université de Poitiers, March 1971.

[30] Mielke, A. Saint-Venant's problem and semi-inverse solutions in nonlinear elasticity. *Arch. Rational Mech. Anal.*, **102** (1988), 205–229.

[31] Mielke, A. Normal hyperbolicity of center manifolds and Saint-Venant's principle. *Arch. Rational Mech. Anal.*, **110** (1990), 353–372.

[32] von Mises, R. On Saint-Venant's Principle. *Bull. Amer. Math. Soc.*, **51** (1945), 555–562.

[33] Neapolitan, R.E. and Edelstein, W.S. Further study of Saint-Venant's principle in linear viscoelasticity. *Z. Angew. Math. Phys.*, **24** (1973), 283–837.

[34] Neapolitan R.E. and Edelstein, W.S. A priori bounds for the secondary boundary value problem in linear viscoelasticity. *SIAM J. Appl. Math.*, **28** (1975), 559–564.

[35] Oleinik, O.A. and Yosifian, G.A. An analogue of Saint-Venant's principle and the uniqueness of solutions of boundary-value problems for parabolic equations in unbounded domains. *Russian Math. Surveys*, **31** (1976), 153–178.

[36] Oleinik, O.A. and Yosifian, G.A. On the asymptotic behaviour at infinity of solutions in linear elasticity. *Arch. Rational Mech. Anal.*, **78** (1982), 29–53.

[37] Payne, L.E. and Weinberger, H.F. Note on a lemma by Finn and Gilbarg. *Acta Math.*, **98** (1957), 297–299.

[38] Barré de Saint-Venant, A.-J.-C. Mémoire sur la tension des prismes, avec des considérations sur leur flexion. *Mém. Divers Savants.*, **14** (1855), 233–560.

[39] Sternberg, E. On Saint-Venant's Principle. *Quart. Appl. Math.*, **11** (1954), 393–402.

[40] Sternberg, E. and Al Khozaie S.A. On Green's function and Saint-Venant's principle in the linear theory of viscoelasticity. *Arch. Rational Mech. Anal.*, **15** (1964), 112–146.

[41] Synge, J.L. The problem of Saint-Venant for a cylinder with free sides. *Quart. Appl. Math.*, **2** (1945), 307–317.

[42] Toupin, R.A. Saint-Venant's principle. *Arch. Rational Mech. Anal.*, **18** (1965), 83–96.

[43] Zanaboni, O. Dimostrazione generale del Principio del de Saint-Venant. *Atti Accad. Naz. Lincei Rend.*, **25** (1937), 117–120.

7
Some Results on Modifications of Three-Dimensional Navier–Stokes Equations

Olga Ladyzhenskaya

In the mid-1960s I suggested a number of modifications to the Navier–Stokes equations (MNS) for the description of the dynamics of viscous fluids when velocity gradients are large [1], see also [2]–[4]. Here we are going to consider those which have the following form:

$$Lv(t) \equiv v_t(t) + v_k(t)v_{x_k}(t) - \text{div } T(\hat{v}(t)) = -\nabla p(t) + f(t), \quad (7.0.1)$$

where $v = (v_1, \ldots, v_n)$ is the velocity field, $\hat{v} = (v_{ij}),\ i, j = \overline{1, n}$, is the matrix with elements $v_{ij} = v_{ix_j} + v_{jx_i})$, $T(\hat{v}) = (T_{ij}(\hat{v}))$ is a symmetric stress tensor and p is the pressure. The space variable $x = (x_1, \ldots, x_n)$ (which was not shown above explicitly in (7.0.1)) is changing in domain Ω of the Euclidian space $\mathbb{R}^n$ ($n = 2$ or 3) and $t \in \mathbb{R}^+ = [0, \infty)$.

7.1. The Three-Dimensional Case ($n = 3$)

For brevity we restrict our considerations by bounded domains Ω. Here we are interested only in the cases when (7.0.1) ensures deterministic description of the dynamics, i.e., in cases when we can guarantee the uniqueness of the principal (first) initial–boundary value problem

$$Lv = -\nabla p + f, \qquad \text{div } v = 0, \qquad v|_{\partial\Omega} = 0, \qquad v|_{t=0} = \varphi. \quad (7.1.1)$$

In [2] we have imposed on $T(\hat{v})$ only the following requirements:

(a) $|T_{ij}(\hat{v})| \leq c(1 + |\hat{v}|^{2\mu})|\hat{v}|, \qquad |\hat{v}| \equiv \left(\sum_{i,j=1}^{n} v_{ij}^2\right)^{1/2};$

(b) tensor $T(\hat{v})$ is symmetric and its elements are continuous functions of $\hat{v}$;

(c) $T_{ij}(\hat{v})v_{ij} \geq (\nu_0 + \nu_1|\hat{v}|^{2\mu})\hat{v}^2, \ \ \hat{v}^2 \equiv |\hat{v}|^2, \quad \nu_k = \text{const} > 0, \ \ k = 0, 1;$

(d) $\displaystyle\int_\Omega [T_{ij}(\hat{v}') - T_{ij}(\hat{v}'')]\,(v'_{ij} - v''_{ij})\,dx \geq \nu_2 \int_\Omega \sum_{i,j=1}^{n} (v'_{ij} - v''_{ij})^2\,dx,$

$\nu_2 = \text{const} > 0,$

for arbitrary v' and v'' from $W^1_{2+2\mu}(\Omega)$, $\mu \geq 0$, which are equal on the boundary $\partial\Omega$.

For such T, the following theorem was proved in [2].

Theorem 7.1. *If T satisfies in $\Omega \in \mathbb{R}^3$ conditions* (a)–(d) *with $\mu \geq \frac{1}{4}$, then problem* (7.1.1) *with $f \in L_{2,1}(Q_T)$, $Q_T = \Omega \times (0, T)$, and $\varphi \in \overset{\circ}{J}(\Omega)$, has a unique solution v belonging to $C([0, T]; \overset{\circ}{J}(\Omega))$ with $v_x \in L_q(Q_T)$, $q \equiv 2 + 2\mu$.*

We will not describe the other properties of solution v proved in [2]. They deal with the smoothness of v with respect to t requiring the introduction of additional spaces. All results in [2] were formulated and proved simultaneously for any domain Ω in $\mathbb{R}^3$, bounded and unbounded. The assumption that Ω is bounded permits some simplifications in the analysis and description of the results. They are caused by the boundedness of the embeddings $W^{l_1}_{q_1}(\Omega) \subset W^{l_2}_{q_2}(\Omega)$ for $q_1 \geq q_2$, $l_1 \geq l_2$. We can use, for example, the following norms $\|u\|_{\overset{\circ}{W}{}^l_q(\Omega)} = \|D^l_x u\|_{q,\Omega}$ in Sobolev space $\overset{\circ}{W}{}^l_q(\Omega)$, where $\|v\|_{q,\Omega}$ is the standard norm in $L_q(\Omega)$ ($\equiv W^0_q(\Omega)$).

The space $\overset{\circ}{J}{}^l_q(\Omega)$ is the closure in the norm of $\overset{\circ}{W}{}^l_q(\Omega)$ of the set

$$\overset{\circ}{J}{}^\infty(\Omega) = \{u \,\|\, u \in C^\infty(\Omega),\ \operatorname{div} u = 0,\ \operatorname{supp} u \subset\subset \Omega\}.$$

$\overset{\circ}{J}{}^l_q(\Omega)$, with $l = 0$ and $q = 2$, we will denote by $\overset{\circ}{J}(\Omega)$ or H_0. For the norm and inner product in $L_2(\Omega)$ we use the notations $\|\cdot\|$ and $(\cdot)$, respectively. It is known that the space $G(\Omega) \equiv \{\nabla p \,\|\, p \in W^1_2(\Omega)\}$ is the orthogonal complement of H_0 in $L_2(\Omega)$ (see [4]). The standard integration by parts gives the identity

$$\int_\Omega \hat{u}^2\,dx = 2\int_\Omega u_x^2\,dx + 2\int_\Omega (\operatorname{div} u)^2\,dx \tag{7.1.2}$$

for any $u \in \overset{\circ}{W}{}^1_2(\Omega)$, so that the norms $\|u\|_{\overset{\circ}{W}{}^1_2(\Omega)} = \|u_x\|$ and $\|\hat{u}\|$ are equivalent in $\overset{\circ}{W}{}^1_2(\Omega)$. The same is true in $\overset{\circ}{W}{}^1_q(\Omega)$ with any $q > 1$. In particular, the inequalities

$$\|u_x\|_{q,\Omega} \leq c_q \|\hat{u}\|_{q,\Omega}, \qquad q > 1, \tag{7.1.3}$$

are true for any $u \in \overset{\circ}{J}{}^1_q(\Omega)$ (see [5]). Due to (7.1.2) the norms $\|u_x\|_{q,\Omega}$ and $\|\hat{u}\|_{q,\Omega}$ are equivalent in $\overset{\circ}{J}{}^1_q(\Omega)$.

Supplement to Theorem 7.1. *All the statements of Theorem* 7.1 *are true for $f = f' + f''$, $f' \in L_{2,1}(Q_T)$, and $f'' = \sum_{i=1}^n \partial_{x_i} f^{(i)}$, $f^{(i)} \in L_{q'}(Q_T)$, $q' = q/(q-1) = (2+2\mu)/(1+2\mu)$. This requirement on f'' implies that the integral*

$$l_{f''}(\eta) \equiv -\int_{Q_T} \sum_{i=1}^{n} f^{(i)}(x,t)\,\eta_{x_i}(x,t)\,dx\,dt, \tag{7.1.4}$$

which formally is equal to $\int_{Q_T} f''(x,t)\eta(x,t)\,dx\,dt$, *is finite for any* $\eta \in L_q(0,T;\overset{\circ}{W}{}^1_q(\Omega))$, $q = 2 + 2\mu$, *and*

$$|l_{f''}(\eta)| \le \sum_{i=1}^{n} \|f^{(i)}\|_{q',Q_T} \|\eta_{x_i}\|_{q,Q_T}. \tag{7.1.5}$$

Moreover, it is sufficient to have (7.1.5) *only for* $\eta \in L_q((0,T);\ \overset{\circ}{J}{}^1_q(\Omega))$.

The possibility of adding such f'' to f is based on the fact that we have used, in the proof of Theorem 7.1 only, the inequality (7.1.5). The source of principal estimates on which the proof of Theorem 7.1 is based is the "energy relation." We get this as a result of inner multiplication of the system (7.0.1) by $v(t)$ and some elementary transformations with regard to other equations of (7.1.1). It has the form

$$\frac{1}{2}\frac{d}{dt}\|v(t)\|^2 + \frac{1}{2}(T_{ij}(\hat{v}(t)),\ v_{ij}(t)) = (f(t),\ v(t)). \tag{7.1.6}$$

By virtue of (b) the following inequality may be deduced from (7.1.6):

$$\frac{d}{dt}\|v(t)\|^2 + \nu_0\|\hat{v}(t)\|^2 + \nu_1\|\hat{v}(t)\|^q_{q,\Omega} \le 2(f(t),\ v(t)), \quad q = 2 + 2\mu. \tag{7.1.7}$$

We estimate the right-hand side of (7.1.7) in different ways depending on our hypothesis concerning f and our aims. We used three versions:

$$j_1(t) \equiv |(f(t),\ v(t))| \le \|f(t)\|\ \|v(t)\|, \tag{7.1.8$_1$}$$

$$j_1(t) \le \|f(t)\|_{(-1)}\|v_x(t)\|, \tag{7.1.8$_2$}$$

$$j_1(t) \le c_{f(t)}\|v_x(t)\|_{q,\Omega}, \tag{7.1.8$_3$}$$

where

$$\|f(t)\|_{(-1)} \equiv \sup\frac{|(f(t),\eta)|}{\|\eta_x\|} \quad \text{for all} \quad \eta \in \overset{\circ}{J}{}^1_2(\Omega),\ [\text{and}]$$

$$c_{f(t)} = \sup\frac{|(f(t),\eta)|}{\|\eta_x\|_{q,\Omega}} \quad \text{for all} \quad \eta \in \overset{\circ}{J}{}^1_q(\Omega).$$

The estimate (7.1.8$_2$) is used if $\nu_0 > 0$ and we do not study the limit $\nu_0 \to 0$. The estimate (7.1.8$_3$) is profitable if $\nu_1 > 0$ and we do not study the limit $\nu_1 \to 0$. The estimate (7.1.8$_1$) is the most restrictive: $f(t) \in L_2(\Omega)$. In Theorem 7.1 we used (7.1.8$_1$) and in the Supplement to Theorem 7.1 it is necessary to use (7.1.8$_3$). In what follows we restrict ourselves to the hypothesis that $f \in L_2(Q_T)$ and that estimate (7.1.8$_1$) is used.

We have described in [2] not only how to prove the solution $v(t) \equiv v(t, \varphi)$ to problem (7.1.1) its continuity in H_0 with respect to $t \in [0, T]$ but also the following facts:

(1) $\|v(t)\|^2$ is absolutely continuous with respect to t and the energy relation (7.1.6) is true for almost all $t \in [0, T]$;

(2) for the difference $u(t) \equiv v'(t) - v''(t)$ of two solutions to (7.1.1) with initial data φ' and φ'', respectively, the relation

$$\begin{aligned} &\frac{1}{2}\frac{d}{dt}\|u(t)\|^2 + \frac{1}{2}(T_{ij}(\hat{v}'(t)) - T_{ij}(\hat{v}''(t)),\ u_{ij}(t)) \\ &\quad = -(u_k(t)\, v'_{x_k}(t) + v''_k(t)\, u_{x_k}(t),\ u(t)) = -(u_k(t)\, v'_{x_k}(t),\ u(t)) \\ &\quad = -\frac{1}{2}(u_k(t)\, u_i(t),\ v'_{ki}(t)) \end{aligned} \tag{7.1.8}$$

holds.

Due to (a)–(c) the following priori estimates

$$\sup_{\tau\in[0,t]} \|v(\tau)\|^2 + \nu_0\|\hat{v}\|^2_{2,Q_t} + \nu_1\|\hat{v}\|^q_{q,Q_t} \le \Phi_1(t, \|v(0)\|), \quad t \in [0, T], \tag{7.1.9}$$

are derived from (7.1.7). Here $\Phi_1(\cdot\,,\,\cdot)$ is a nondecreasing continuous function of its arguments (we do not indicate the dependence Φ_1 on f, ν_k, Ω) which can be calculated explicitly.

Let us single out the case when f does not depend on t and $f \in L_2(\Omega)$. Then the inequality

$$\frac{1}{2}\frac{d}{dt}\|v(t)\|^2 + \nu_0\lambda_1\|v(t)\|^2 \le \|f\|\,\|v(t)\| \tag{7.1.10}$$

can be proved, where λ_1 is the first eigenvalue of the Stokes operator $-\tilde{\Delta} = -P\Delta$ under the Dirichlet condition (see [4, Chap. II]). From (7.1.10) it follows that

$$\|v(t)\| \le \|v(0)\|\, e^{-\nu_0\lambda_1 t} + (1 - e^{-\nu_0\lambda_1 t})\frac{\|f\|}{\nu_0\lambda_1}. \tag{7.1.11}$$

From (7.1.11) and (7.1.7) we can deduce

$$\sup_{t\in\mathbb{R}^+} \int_t^{t+1} (\nu_0\|\hat{v}(\tau)\|^2 + \nu_1\|\hat{v}(\tau)\|^q_{q,\Omega})\, d\tau \le \Phi_2(\|v(0)\|). \tag{7.1.12}$$

The uniqueness theorem in the class of generalized solutions described in Theorem 7.1 is derived with the help of (7.1.8). In a similar way, the continuous dependence of solutions $v(t, \varphi)$ in the topology of H_0 is also proved. Namely, from (7.1.8) and (d) it follows that

$$\begin{aligned} \frac{d}{dt}\|u(t)\|^2 + \nu_2\|\hat{u}(t)\|^2 &\le |(u_k(t)\, u_i(t),\ v'_{ki}(t))| \le \|u(t)\|^2_{2q',\Omega}\|\hat{v}'(t)\|_{q,\Omega} \\ &\le \beta^2\|u_x(t)\|^{2\alpha}\|u(t)\|^{2(1-\alpha)}\|\hat{v}'(t)\|_{q,\Omega} \le \nu_2\|u_x(t)\|^2 \\ &\quad + c\|\hat{v}'(t)\|^{1/(1-\alpha)}_{q,\Omega}\|u(t)\|^2, \qquad \alpha = \frac{3}{2q}. \end{aligned} \tag{7.1.13}$$

Here we have used the multiplicative inequality

$$\|u\|_{p,\Omega} \leq \beta \|u_x\|^{\alpha}_{m,\Omega} \|u\|^{1-\alpha}_{r,\Omega}, \qquad \alpha = \left(\frac{1}{r} - \frac{1}{p}\right)\left(\frac{1}{r} - \frac{1}{m} + \frac{1}{n}\right)^{-1}, \tag{7.1.14}$$

that holds for any $u \in \overset{\circ}{W}{}^1_m(\Omega)$, $m \geq 1$, $r \geq 1$, in the arbitrary domain $\Omega \subset \mathbb{R}^n$. In (7.1.14) for $m < n$ the number p can take any value from $[r, \bar{m}]$, where $\bar{m} = nm/(n-m)$, if $r \leq \bar{m}$, and any value from $[\bar{m}, r]$, if $r \geq \bar{m}$, and $\beta = (\bar{m}(n-1)/n))^{\alpha}$. When p varies between r and $\bar{m}$ the number α changes between 0 and 1 at both ends; for $r = p = \bar{m}$ any number from [0, 1] can be taken as α. If $m \geq n$ the number p can be arbitrary from $[r, \infty)$ and for $m > n$ (7.1.14) holds even for $p = \infty$ with $\alpha = mn/[mn - r(m-n)]$. The constant β depends on the numerical parameters n, m, p, r but does not depend on u and Ω (see [6, Chap. II]).

To derive (7.1.13) we have taken (7.1.14) with $p = 2q'$, $m = 2$, $r = 2$, and $n = 3$. We can integrate (7.1.13) if $1/(1-\alpha) \leq q$, i.e., if $\mu \geq \frac{1}{4}$. In this case, we have

$$\frac{d}{dt}\|u(t)\|^2 + \nu_2 \|u_x(t)\|^2 \leq c\, \|\hat{v}'(t)\|^{1/(1-\alpha)}_{q,\Omega} \|u(t)\|^2$$

and

$$\begin{aligned} \|u(t)\|^2 &\leq \|u(0)\|^2 \exp\left\{c \int_0^t \|\hat{v}'(\tau)\|^{1/(1-\alpha)}_{q,\Omega} d\tau\right\} \\ &\leq \|u(0)\|^2 \exp\left\{ct^{1-(1-\alpha)/q} \times \|\hat{v}'\|^{(1-\alpha)/q}_{q,Q_t}\right\}, \qquad \mu \geq \frac{1}{4}. \end{aligned} \tag{7.1.15}$$

From (7.1.15) follow the uniqueness theorem and the continuous dependence of solutions $v(t, \varphi)$ on φ in the topology of H_0. To sum up our reasoning we can infer that: if f is independent of t and $f \in L_2(\Omega)$, then the family of solution operators $V_t : \varphi \to v(t) \equiv v(t, \varphi)$ of problem (7.1.1) forms a continuous semigroup $\{V_t, t \in \mathbb{R}^+, H_0\}$. For this semigroup, the ball $B_{R_0}(H_0) = \{u \parallel u \in H_0, \|u\| \leq R_0 = \|f\| (\nu_0\lambda_1)^{-1}\}$ proves a B–attracting set and balls $B_R(H_0)$ with $R > R_0$ are B–absorbing sets. All this remains true for $f = f' + \sum_{i=1}^n \partial f^{(i)}/\partial x_i$, $f' \in L_2(\Omega)$, $f^{(i)} \in L_q(\Omega)$.

But the properties of the semigroup $\{V_t, t \in \mathbb{R}^+, H_0\}$ listed above are not sufficient to state the existence of a minimal global B–attractor $\mathcal{M}$. The existence of such an attractor we have proved in [7] for the cases when T has special structure

$$T_{ij}(\hat{v}) = \frac{\partial \mathcal{D}(\hat{v})}{\partial v_{ij}}, \tag{7.1.16}$$

where $\mathcal{D} : M^{3\times3}_{\text{symm}} \to \mathbb{R}^+$ is a potential corresponding to T and having the properties:

(a′) $\mathcal{D}$ is continuous, continuously differentiable, and $\mathcal{D}(0) = 0$;

(b′) $$\left|\frac{\partial \mathcal{D}(\hat{v})}{\partial \hat{v}}\right| \equiv \left[\sum_{i,j=1}^{3}\left(\frac{\partial \mathcal{D}(\hat{v})}{\partial v_{ij}}\right)^2\right]^{1/2} \leq c\,(1+|\hat{v}|^{2\mu})\,|\hat{v}|;$$

(c′) $\mathcal{D}(\hat{v}) \geq \nu_0 \hat{v}^2 + \nu_1 |\hat{v}|^q, \qquad q = 2+2\mu, \quad \nu_0, \nu_1 > 0;$ *and*

(d′) $$\left(\frac{\partial^2 \mathcal{D}(\hat{v})}{\partial \hat{v}^2}\zeta\right) : \zeta \equiv \sum_{i,k,j,l=1}^{3} \frac{\partial^2 \mathcal{D}(\hat{v})}{\partial v_{ik}\partial v_{jl}} \zeta_{ik}\zeta_{jl} \geq \nu_2 \sum_{i,k=1}^{3} \zeta_{ik}^2, \qquad \nu_2 > 0.$$

Such T satisfy all the requirements (a)–(d). So Theorem 7.1 and all the statements made just after it are true. Before we formulate the results of [7] concerning attractors let's remember some results from our first publication [3] on a global unique solvability for MNS. There problem (7.1.1) was investigated for T having the structure (7.1.16) and v_{ij} equal to $v_{ix_j} + v_{jx_i}$ or v_{ix_j}. All the proofs are given for $T_{ij} = (\nu_0 + \nu_1 |v_x|^{2\mu}) v_{ix_j}$ but it was pointed out that they hold for other T_{ij} enumerated in the introduction to [3]. In [1]–[3] some arguments of physical character are given in favor of T considered in [3]. But for the proof of the mathematical statements we used in [3], and not some specific form of $\mathcal{D}$, but only the fact that the sums $T_{ij}(v_x)\,\partial_t v_{ix_j}$ or $T_{ij}(\hat{v})\,\partial_t v_{ij}$ are full derivatives with respect to t of functions generating some "strong" norms. In other words, it is sufficient to have (7.1.16) in [3] (as well as in [7]) with any $\mathcal{D}$ satisfying conditions $(a') - (d')$. In these cases

$$T_{ij}(\hat{v}(x,t))\,\partial_t v_{ij}(x,t) = \partial_t \mathcal{D}(\hat{v}(x,t)). \tag{7.1.17}$$

In its turn, this leads to one more integral relation that helps to obtain the estimates of some strong norms of v. It is obtained as a result of the inner multiplication of equation (7.0.1) by $v_t(t)$ and has the form

$$\|v_t(t)\|^2 + \frac{1}{2}\frac{d}{dt}\|\mathcal{D}(\hat{v}(t))\|_{1,\Omega} = -(v_k(t)\,v_{x_k}(t),\ v_t(t)) + (f(t),\ v_t(t)). \tag{7.1.18}$$

We estimated the term $j_1(t) = -(v_k(t)v_{x_k}(t), v_t(t))$ in [3] in the following way:

$$\begin{aligned} |j_1(t)| &\leq \|v(t)\|_{2+2\mu)/\mu,\Omega}\|v_x(t)\|_{2+2\mu,\Omega}\|v_t(t)\|_{2,\Omega} \\ &\leq \beta\|v_x(t)\|^2_{2+2\mu,\Omega}\|v_t(t)\|_{2,\Omega} \leq \frac{1}{4}\|v_t(t)\|^2 + \beta^2\|v_x(t)\|^4_{2+2\mu,\Omega} \end{aligned} \tag{7.1.19}$$

assuming that $\mu \geq \frac{1}{5}$. In this case, we can derive from (7.1.18), (7.1.19), and (a') the inequality

$$\begin{aligned} &\|v_t(t)\|^2 + \frac{d}{dt}\|\mathcal{D}(\hat{v}(t))\|_{1,\Omega} \leq \\ &\qquad c\,\|v_x(t)\|^{2-2\mu}_{2+2\mu,\Omega}\|\mathcal{D}(\hat{v}(t))\|_{1,\Omega} + 2\|f(t)\|^2. \end{aligned} \tag{7.1.20}$$

If $v(0) \in \overset{\circ}{J}{}^1_{2+2\mu}(\Omega)$ we can integrate (7.1.20) from $t = 0$ to t bearing in mind that (7.1.12) gives a majorant for $\int_0^t \|v_x(\tau)\|^{2+2\mu}_{2+2\mu,\Omega}\,d\tau$. Let us write the result in the form

$$\sup_{\tau\in[0,t]} (\nu_0\|\hat{v}(\tau)\|^2 + \nu_1\|\hat{v}(\tau)\|^q_{q,\Omega}) + \|v_t\|^2_{2,Q_t} \le \Phi_3(t,\ \|\hat{v}(0)\|_{q,\Omega}),$$
$$q = 2+2\mu, \qquad \mu \ge \frac{1}{5}. \tag{7.1.21}$$

On the basis of this estimate in [3] was proved the theorem of existence for problem [2] with $\varphi \in \overset{\circ}{J}{}^1_{2+2\mu}(\Omega)$ and $\mu \ge \frac{1}{5}$ (Theorem 1.1 of [3]). But we did not write there that in this case the uniqueness theorem is also true. Moreover, in the next theorem (Theorem 1.2 of [3]) we asserted a uniqueness for $\mu \ge \frac{1}{4}$. But in Theorem 1.2 we considered a wider class of generalized solutions (the same as in Theorem 7.1) and for this class the restriction $\mu \ge \frac{1}{4}$ may be necessary. Let us improve the deficiency of [3] and formulate the following:

Theorem 7.2. *Problem* (7.1.1) *has for* $\varphi \in \overset{\circ}{J}{}^1_{2+2\mu}(\Omega)$, $f \in L_2(Q_T)$, $\mu \ge \frac{1}{5}$, *a unique solution* v *having the properties indicated in Theorem* 7.1 *and also the properties* $v \in L_\infty((0,T);\ \overset{\circ}{J}{}^1_{2+2\mu}(\Omega))$, $v_t \in L_2(Q_T)$. *The uniqueness in such a class of solutions holds for arbitrary* $\mu \ge 0$.

We find the solutions v of (7.1.1) as limits of the Galerkin–Faedo approximations $v^m(t)$, $m = 1, 2, \ldots$. The relations (7.1.5) and (7.1.18) hold for any $v^m(t)$, and therefore the estimates (7.1.9) and (7.1.21) for $v^m(t)$ are also true. It permits us to pass to the limit and prove Theorem 7.2. The uniqueness follows from (7.1.13) for any $\mu \ge 0$ as now we have the boundedness of $\|\hat{v}'(t)\|_{2,\Omega}$ on $[0,T]$, and can take $\alpha = 3/4$.

We extract useful information from (7.1.20) and for $v(0) \in H_0$. Namely, as in [8], we multiply (7.1.20) by t and rewrite the result in the form

$$t\|v_t(t)\|^2 + \frac{dy(t)}{dt} \le y(t) + c\|v_x(t)\|^{2-2\mu}_{2+2\mu,\Omega}\, y(t) + 2t\,\|f(t)\|^2, \tag{7.1.22}$$

where $y(t) = t\|\mathcal{D}(\hat{v}(t))\|_{1,\Omega}$. Due to (7.1.9) we get from (7.1.22) the estimate

$$t\|\mathcal{D}(\hat{v}(t))\|_{1,\Omega} + \int_0^t \tau\|v_\tau(\tau)\|^2\,d\tau \le \Phi_4(t,\ \|v(0)\|). \tag{7.1.23}$$

The right-hand side of (7.1.23) depends only on the "weak" norm of $v(0)$, and on the left-hand side of (7.1.23) we have a stronger norm of $v(t)$ when $t > 0$. It guarantees the compactness of $V_t : H_0 \to H_0$ for $t > 0$ and gives the possibility of following [8] or using the results about attractors for semigroups of the first class [9], [10] and obtain:

Theorem 7.3. *Let* $T(\hat{v})$ *have the form* (7.1.16) *with* $\mathcal{D}(\hat{v})$ *satisfying the conditions* $(a') - (d')$ *with* $\mu \ge \frac{1}{4}$ *and let* f *belong to* H_0 *and not depending on* t. *Then the solution operators* V_t *of problem* (7.1.1) *form a continuous semigroup* $\{V_t,\ t \in \mathbb{R}^+,\ H_0\}$ *of the first class. The balls* $B_R(H_0)$ *with* $R > R_0 = \|f\|/\nu_0\lambda_1$ *are B–absorbing sets for the semigroup. It has a minimal global B–attractor* $\mathcal{M}$

which is an invariant compact subset of H_0 *bounded in the space* $\overset{\circ}{J}{}^1_{2+2\mu}(\Omega)$. *There is a common majorant* c *for*

$$\sup_{\varphi\in\mathcal{M}}\sup_{t\in\mathbb{R}^1}\left\{\|\varphi_x\|_{2+2\mu,\Omega},\ \int_t^{t+1}\|v_\tau(\tau,\varphi)\|^2\,d\tau\right\}\le c. \tag{7.1.24}$$

The number of determining modes on $\mathcal{M}$ *is finite.*

As we have explained, Theorem 7.3 follows from [2] and [8]. It is also contained in the statements of [7]. Since [7] is not yet published, we shall take from [7] one result which generalizes Theorem 7.3.

Theorem 7.4. *Let all the conditions of Theorem 7.3 be fulfilled, except the condition* $f\in H_0$, *which will be replaced by* $f\in \overset{\circ}{J}{}_2^{(-1)}(\Omega)$. *Then all the statements of Theorem 7.3 are true (with some other majorants in estimates). Moreover, the trajectories* $v(t,\varphi)$ *on* $\mathcal{M}$ *have the derivatives* $v_{xt}\in L_2(\Omega\times(-T,T))$, $\forall T$, *and there is a common majorant* c' *for*

$$\sup_{\varphi\in\mathcal{M}}\sup_{t\in\mathbb{R}^1}\left\{\|v_t(t,\varphi)\|,\ \int_t^{t+1}\|v_{\tau x}(\tau,\varphi)\|^2\,d\tau\right\}\le c'. \tag{7.1.25}$$

To prove Theorem 7.4 we involve one more relation—the inner product of $v_t(t)$ and the equation that is obtained from (7.0.1) by differentiation with respect to t. After elementary transformations it has the form

$$\frac{1}{2}\frac{d}{dt}\|v_t(t)\|^2+\frac{1}{2}\int_\Omega\frac{\partial^2\mathcal{D}(\hat v(x,t))}{\partial v_{ij}\partial v_{kl}}v_{ijt}(x,t)$$
$$\times\, v_{klt}(x,t)\,dx=-(v_{kt}(t)\,v_{x_k}(t),\,v_t(t))+(f_t(t),\,v_t(t)). \tag{7.1.26}$$

The last term in (7.1.26) is equal to zero (as f does not depend on t) and the terms $(f,\ v(t))$ and $(f,\ v_t(t))$ in (7.1.6) and (7.1.18) we majorize in the following way:

$$|(f,\ v(t))|\le\|f\|_{\overset{\circ}{J}{}_2^{(-1)}(\Omega)}\|v_x(t)\|\equiv c_f\|v_x(t)\|, \tag{7.1.27}$$

$$|(f,\ v_t(t))|\le c_f\|v_{xt}(t)\|. \tag{7.1.27$_1$}$$

The finiteness of c_f is our requirement concerning f in Theorem 7.4.

We manipulate with (7.1.6), (7.1.18), and (7.1.26) in [7] in the same manner as we did in [11] and [12] with analogous relations for the two-dimensional Navier–Stokes equations. For f independent on t and for f having $f_t\in L_2(0,T;\ \overset{\circ}{J}{}_2^{(-1)}(\Omega))$ we derive from (7.1.6), (7.1.18), and (7.1.26) an instant smoothing of solutions $v(t,\varphi)$, $\varphi\in H_0$, to problem (7.1.1) and the existence of derivatives $v_x(t,\varphi)$, $v_t(t,\varphi)$, $v_{xt}(t,\varphi)$ for $t>0$.

Let us emphasize that we do all our conclusions using not system (7.0.1), but only the relations (7.1.6), (7.1.18), and (7.1.26) and the integral identity

$$(v_t(t),\ \eta) + \frac{1}{2}(T_{ij}(\hat{v}(t),\ \eta_{ij}) + (v_k(t)v_{x_k}(t), \eta) = (f(t),\ \eta),$$
$$\forall \eta \in \dot{J}^{\infty}(\Omega). \tag{7.1.28}$$

All these relations hold for Galerkin–Faedo approximations with the arbitrary fundamental system in $\overset{\circ}{J}{}^1_{2+2\mu}(\Omega)$. They, or their analogues, are true also for some other approximations. This permits us to prove for (7.1.1) a part of the results which we have proved in [8] and [9] for the Navier–Stokes equations. But there are open questions concerning systems (7.0.1). Let us formulate some of them.

(1) Are the solution operators V_t of the problem (7.1.1) differentiable on some phase space where the global unique solvability takes place?

In most of the theorems concerning the finiteness of dimensions of sets $\mathcal{A}$ which are invariant with respect to an operator V there is the requirement that V is uniformly differentiable on $\mathcal{A}$ (see [10, and references therein]). The other conditions of these theorems concern the linearizations of equation (7.0.1) on its solutions $v(t, \varphi)$ belonging to $\mathcal{A}$. Checking these requirements in all the cases I know uses the boundedness of coefficients in principal terms of the linearizations. Such boundedness requires the existence of a majorant $c < \infty$ for

$$\sup_{\varphi\in\mathcal{A}} \max_{x\in\Omega} |\hat{v}(x, t;\ \varphi)| \leq c. \tag{7.1.29}$$

We have proved the boundedness of the norms indicated in (7.1.24) and (7.1.25). But it does not give (7.1.29). The estimate (7.1.29) would also help to prove the differentiability of V_t on $\mathcal{A}$.

In connection with this it is important to find cases when (7.1.29) takes place. In paper [13] the a priori estimate

$$j(t) \equiv \int_{Q_t} (1 + |\hat{v}|^{2\mu})\, |\hat{v}_x|^2\, dx\, dt \leq \Phi(t,\ \|\hat{v}(0)\|_{9,\Omega}), \quad 9 = 2 + 2\mu \tag{7.1.30}$$

for solutions v of the system (7.0.1) submitted to periodical boundary conditions (Ω in this case is a cube or a parallelepiped in $\mathbb{R}^3$). It has been done for $\mathcal{D}$ satisfying the conditions (a')–(c') with $\mu \geq \frac{1}{10}$ and condition

$$(\text{d}'') \quad \left(\frac{\partial^2\mathcal{D}(\hat{v})}{\partial \hat{v}^2}\zeta\right) : \zeta \geq \nu_2(1 + |\hat{v}|^{2\mu}) \sum_{i,k=1}^{3} \zeta_{ik}, \qquad \nu_2 > 0.$$

The additional information (7.1.30) has permitted the authors of [13] to prove the unique solvability of the system (7.0.1) with periodic boundary conditions in the space $\tilde{J}^1_9(\Omega)$ for $\mu \geq \frac{1}{10}$. (Here $\tilde{J}^1_9(\Omega)$ is the space of divergent-free vector fields satisfying periodic boundary conditions and having a norm equivalent to the norm of $W_9^1(\Omega)$.)[1] From the results of [13] it follows that for f independent of t the solution operators V_t form a continuous semigroup $\{V_t,\ t \in \mathbb{R}^+,\ \tilde{J}_9^1(\Omega)\}$.

There is the question: Has this semigroup a minimal global B–attractor $\mathcal{M}$? Remember that such $\mathcal{M}$ has to attract any bounded subset on $X \equiv \tilde{J}_9^1(\Omega)$ in the topology of X (but not in a weaker topology). For $\mu \geq \frac{1}{4}$ Theorems 7.3 and 7.4 of this chapter guarantee the existence of such an attractor in the phase space $X \equiv \overset{\circ}{J}(\Omega)$. It is true, of course, for periodic boundary conditions too.

(2) To find $T(\hat{v})$ for which (7.0.1) with f independent of t and with at least periodic boundary conditions has a global attractor with a finite Hausdorff dimension.

Among MNS equations suggested by us in [1–3] there are equations with finite-dimensional global attractors. To these equations belong, for example, equations (7.0.1) with

$$T_{ij}(\hat{v}) = \nu(\hat{v})\, v_{ij}, \qquad \nu(\hat{v}) = \nu_0 + \nu_1 \|\hat{v}\|^2, \qquad \nu_k = \text{const} > 0. \tag{7.1.31}$$

For these in [3] was proved the global unique solvability in the phase spaces $\overset{\circ}{J}(\Omega)$ and $W_2^2(\Omega) \cap \overset{\circ}{J}{}_2^1(\Omega)$ (under the corresponding hypotheses on φ and f). It is also true in the phase space $\overset{\circ}{J}{}_2^1(\Omega)$. In [15] it is proved that if Ω is a bounded domain in $\mathbb{R}^3$ with a smooth boundary $\partial\Omega$ and f belonging to $L_2(\Omega)$ and does not depend on t, then solution operators V_t form a continuous semigroup $\{V_t,\ t \in \mathbb{R}^+,\ \overset{\circ}{J}(\Omega)\}$ of the first class (in the terminology of [9] or [10]) having bounded B–absorbing sets. From this it follows ([9], [10]) that the semigroup has a compact minimal B–attractor $\mathcal{M}$. In [15] were proved analogues of all the properties stated in [8] for the two-dimensional Navier–Stokes equations. Among them the unique continuation of the semigroup $\{V_t,\ t \in \mathbb{R}^+,\ \mathcal{M}\}$ to a continuous group $\{V_t,\ t \in \mathbb{R}^1,\ \mathcal{M}\}$ was proved. The following theorem also holds:

Theorem 7.5. *Let* $T(\hat{v})$ *in* (7.0.1) *have the form* (7.1.31), *let* f *belong to* $L_2(\Omega)$ *and be independent on* t, *and let* Ω *be a bounded domain in* $\mathbb{R}^3$ *with* $\partial\Omega \subset C^2$. *Then problem* (7.1.1) *is globally uniquely solvable in* $X = \overset{\circ}{J}{}_2^1(\Omega)$, *and its solution operators* V_t *form a continuous semigroup* $\{V_t,\ t \in \mathbb{R}^+,\ \overset{\circ}{J}{}_2^1(\Omega)\}$. *It has a compact minimal global* B*–attractor* $\mathcal{M}$. *The operators* V_t *are uniformly differentiable on* $\mathcal{M}$, *and* $\mathcal{M}$ *has a finite fractal dimension in* X.

The proofs of all the statements of Theorem 7.5 except the two last are analogous to the proofs of [15]. The principal line of the proof of the differentiability of V_t

---

1. The possibility of reducing the restriction on μ until $\mu \geq \frac{1}{10}$ was declared in [14]. Namely, in Theorem 4.2 of [14] it is asserted that the global unique solvability of the problem (7.1.1) in any $\Omega \subset \mathbb{R}^3$ for arbitrary $\varphi \in \overset{\circ}{J}(\Omega)$ and $f \in L_2(Q_T)$. The solutions v have, principally, the same properties as in Theorem 7.1. But in [14] there are no explanations as to what is possible for the reduction of the requirement $\mu \geq \frac{1}{4}$ to the requirement $\mu \geq \frac{1}{10}$.

on $\mathcal{M}$ is the same as for the two-dimensional Navier–Stokes equations, but its realization is more complicated because of the nonlinearity of the principal term $\nu(\hat{v})\,\Delta v$ of equation (7.0.1). And what is more, for this reason, we could not prove the differentiability of V_t in $\overset{\circ}{J}(\Omega)$ and passed to phase space $\overset{\circ}{J}{}_2^1(\Omega)$. For the estimation of the fractal dimension of $\mathcal{M}$ we have used Theorems 4.8 and 4.9 of [10] and verified that their conditions are fulfilled in the phase space $X = \overset{\circ}{J}{}_2^1(\Omega)$.

7.2. The Two-Dimensional Problems ($n = 2$)

Let all the data in problem (7.1.1) not depend on the variable x_3 and we study the problem in $\Omega \subset \mathbb{R}^2$. Let us enumerate analogues to the theorems of the previous section which hold for the bounded domain $\Omega \subset \mathbb{R}^2$.

Theorem 7.1 and the Supplement to Theorem 7.1 are true for any $\mu \geq 0$.

Theorem 7.2 is true for any $\mu \geq 0$.

Theorems 7.3 and 7.4 hold for any $\mu \geq 0$.

Theorem 7.4 for $\mu = 0$ (and for smooth and nonsmooth potentials $\mathcal{D}$) is proved in [16]. A short account of [16] is published in [17].

The unique solvability of problem (7.1.1) in classes of functions which have the smoothness described in Theorems 7.3 and 7.4 has been proved also for f depending on t. Such theorems for $n = 2$ are in [16].

Theorem 7.5 holds for $n = 2$.

This paper is based on material reported at the meeting in Ferrara (October 1992) in honor of James B. Serrin.

References

[1] Ladyzhenskaya, O.A. *On Nonlinear Problems of Continuous Mechanics*. Trudy of International Mathematical Congress 1966, Moscow, 1968, pp. 560–572.

[2] Ladyzhenskaya, O.A. On some modifications of the Navier–Stokes equations for large gradients of velocity. *Zap. Nauchn. Sem., LOMI*, **7** (1968), 126–154.

[3] Ladyzhenskaya, O.A. On some new equations for the description of dynamics of incompressible fluids and on a global solvability for these equations the boundary value problems. *Trudy Steklov Math. Inst.* **102** (1967), 85–104.

[4] Ladyzhenskaya, O.A. *Mathematical problems of viscous incompressible fluids*. First Russian edition, 1961, Nauka, Moscow. Second Russian edition, 1970, Nauka, Moscow.

[5] Nečas, J. *Sur les Norms Équivalentes dans $W_p^k(\Omega)$ et sur la Coercitivé des Formes Formellemant positives*. Les Presses de l'Université de Montréal, 1966.

[6] Ladyzhenskaya, O.A. and Uraltseva, N.N. *Linear and Quasilinear Elliptic Equations*, 1973, 2nd edn. Nauka, Moscow.

[7] Ladyzhenskaya, O.A. Attractors for the modifications of the three-dimensional Navier–Stokes equations. *Philos. Trans. Roy. Soc. Ser.* A, **346** (1994), pp. 173–190.

[8] Ladyzhenskaya, O.A. On the dynamical system generated by the Navier–Stokes equations. *Zap. Nauchn. Sem. LOMI*, **27** (1972), 91–114.

[9] Ladyzhenskaya, O.A. On finding the minimal global attractors for the Navier–Stokes equations and other PDE. *Uspehi Mat. Nauk*, **42**, no. 6 (1987), 25–60.

[10] Ladyzhenskaya, O.A. *Attractors for Semigroups and Evolution Equations*. Lezioni Lincee, Roma, 1988. Cambridge University Press, Cambridge, 1991.

[11] Ladyzhenskaya, O.A. First boundary value problem for the Navier–Stokes equations in domains with nonsmooth boundaries. *C. R. Acad. Sci. Paris*, **314**, Série I (1992), 253–258.

[12] Ladyzhenskaya, O.A. Some globally stable approximations for the Navier–Stokes equations and some other equations for viscous incompressible fluids. *C. R. Acad. Sci. Paris*, **315**, Série I (1992), 387–392.

[13] Málek, J., Nečas, J., and Růžička, M. On the non-Newtonian incompressible fluids. *Math. Models Methods Appl. Sci.*, **3**, no. 1 (1993), 35–63.

[14] Du, Q. and Gunzburger, M.D. Analysis of a Ladyzhenskaya model for incompressible viscous flow. *J. Math. Anal. Appl.*, **155** (1991), 21–45.

[15] Ladyzhenskaya, O.A. On limit states for some modifications of the Navier–Stokes equations in the three-dimensional space. *Zap. Nauchn. Sem. LOMI*, **84** (1979), 131–146.

[16] Ladyzhenskaya, O.A. and Seregin G. On semigroups generated by initial-boundary value problems describing two-dimensional visco-plastic flows. In the book "Nonlinear Evolution Equations," American Math. Society Translations, Ser. 2, 164, p 90–124. Advances in the Mathematical Sciences-22, 1995.

[17] Ladyzhenskaya, O.A. and Seregin G. On a global stability of the two-dimensional visco-plastic flows. *Jyväskylä–St. Petersburg Seminar on Partial Differential Equations and Numerical Methods*. University of Jyväskylä, Department of Mathematics, Report 56, 1993, pp. 43–52.

8
An Integral Equation in Probability

J.B. McLeod

8.1. Introduction

The integral equation

$$f(x) = \frac{1}{\sqrt{2\pi}} \int_{-\infty}^{\infty} e^{-x^2/2(\beta_0+\beta_1 v^2)} \frac{f(v)}{\sqrt{\beta_0 + \beta_1 v^2}}\, dv,$$

where the quantities β_0 and β_1 are positive constants, makes its appearance in the discussion of ARCH models in statistics. The original model was proposed by Engle [1], and Pantula [2] was the first to write down the equation in this connection. The function f is a distribution function and as such is to have the properties that f is nonnegative and that its integral is unity. The questions of interest (apart from existence and uniqueness of a solution) are how the number of moments depends on the parameter β_1 and what is the asymptotic behavior of $f(x)$ as $|x| \to \infty$.

These questions are answered in the theorem that follows (and in its proof). I am much indebted to Professor Satish Iyengar for bringing this problem to my attention.

Theorem. (i) *If β_1 is such that, given $s > 1$,*

$$(2\beta_1)^{(s-1)/2}\Gamma(\tfrac{1}{2}s) \quad < \sqrt{\pi}, \tag{8.1.1}$$

then there exists a unique positive solution of

$$f(x) = \frac{1}{\sqrt{2\pi}} \int_{-\infty}^{\infty} e^{-x^2/2(\beta_0+\beta_1 v^2)} \frac{f(v)}{\sqrt{\beta_0 + \beta_1 v^2}} dv \tag{8.1.2}$$

with

$$\int_{-\infty}^{\infty} f(x)\, dx = 1, \tag{8.1.3}$$

and the solution has the property that

$$\int_{-\infty}^{\infty} x^{s-1} f(x)\, dx < \infty. \tag{8.1.4}$$

(ii) *If*

$$(2\beta_1)^{(s-1)/2}\Gamma(\tfrac{1}{2}s) \geq \sqrt{\pi}, \tag{8.1.5}$$

then there exists no positive solution satisfying (8.1.3) *and* (8.1.4).

Remarks. 1. The condition (8.1.3) is merely a normalization. Otherwise the solution to (8.1.2) could not be unique since any multiple of a solution would also be a solution.

2. Clearly any solution to (8.1.2) must be even, and so we can (and will) consider equivalently the equation

$$f(x) = \sqrt{\frac{2}{\pi}} \int_0^\infty e^{-x^2/2(\beta_0+\beta_1 v^2)} \frac{f(v)}{\sqrt{\beta_0 + \beta_1 v^2}}\, dv. \tag{8.1.6}$$

3. The theorem shows that, as β_1 increases, the number of moments decreases. In particular, for the first moment ($s = 2$), the critical value of β_1 is $\beta_1 = \sqrt{\pi/2}$. For the second moment ($s = 3$), the critical value is $\beta_1 = 1$, and so on.

4. By differentiating (8.1.6), we see that any positive solution is necessarily decreasing for $x > 0$.

5. It can be shown and indeed is shown as part of the proof of the theorem, that any positive solution has the property that

$$f(x) \sim Ax^{-s_0} \quad \text{as} \quad x \to \infty,$$

for some positive constant A and where s_0 (given β_1) satisfies

$$(2\beta_1)^{(s_0-1)/2}\Gamma(\tfrac{1}{2}s_0) = \sqrt{\pi}.$$

(The appendix discusses the uniqueness of the value of s_0.)

6. We can remove β_0 from the equation. Setting $x = \sqrt{\beta_0}y$ and $v = \sqrt{\beta_0}w$, we have

$$f(\sqrt{\beta_0}y) = \sqrt{\frac{2}{\pi}} \int_0^\infty e^{-y^2/2(1+\beta_1 w^2)} \frac{f(\sqrt{\beta_0}w)}{\sqrt{1+\beta_1 w^2}}\, dw,$$

so that

$$f(x, \beta_0) = \frac{1}{\sqrt{\beta_0}} f(x/\sqrt{\beta_0},\ 1).$$

(The multiplying factor $1/\sqrt{\beta_0}$ is introduced to ensure (8.1.3).)

7. There are certain easy particular cases. As $\beta_0 \to 0$, then formally, from Remark 6, $f(x, \beta_0)$ for any finite x behaves like $f(x/\sqrt{\beta_0},\ 1)$, and so, from Remark 5, like x^{-s_0}; and it can be verified by direct calculation that indeed x^{-s_0} is a solution of (8.1.2) with $\beta_0 = 0$.

If $\beta_1 = 0$, then $f(x)$ is a multiple of $e^{-x^2/2\beta_0}$.

8. Much of the analysis extends to the problem where $\beta_0 + \beta_1 v^2$ is replaced by some other suitable function $g(v)$. For example, we would suspect that if we replace $\beta_0 + \beta_1 v^2$ by $\beta_0 + \beta_1 |v|^p$, then, for $p < 2$, there is a unique positive solution for all β_1, possessing all moments, while if $p > 2$ there is no positive solution satisfying (8.1.3) and (8.1.4).

8.2. Proof of Part (ii) of the Theorem

Taking the equation in the form (8.1.5), multiplying by x^{s-1}, and integrating, we have

$$\begin{aligned}
\int_0^\infty x^{s-1} f(x)\,dx &= \sqrt{\frac{2}{\pi}} \int_0^\infty \frac{f(v)}{\sqrt{\beta_0 + \beta_1 v^2}}\,dv \int_0^\infty x^{s-1} e^{-x^2/2(\beta_0+\beta_1 v^2)}\,dx \\
&= \sqrt{\frac{2}{\pi}} \int_0^\infty f(v)(\beta_0 + \beta_1 v^2)^{(s-1)/2}\,dv \int_0^\infty y^{s-1} e^{-y^2/2}\,dy \\
&= \sqrt{\frac{2}{\pi}} \left(\int_0^\infty f(v)(\beta_0 + \beta_1 v^2)^{(s-1)/2} dv \right) \int_0^\infty (2z)^{(1/2)(s-2)} e^{-z}\,dz \\
&= 2^{(1/2)(s-2)} \Gamma(\tfrac{1}{2}s) \sqrt{\frac{2}{\pi}} \int_0^\infty f(v)(\beta_0 + \beta_1 v^2)^{(s-1)/2}\,dv \\
&> 2^{(1/2)(s-2)} \Gamma(\tfrac{1}{2}s) \sqrt{\frac{2}{\pi}}\, \beta_1^{(1/2)(s-1)} \int_0^\infty f(v) v^{s-1}\,dv,
\end{aligned}$$

and this is impossible, for solutions satisfying (8.1.4) (unless $f = 0$ which contradicts (8.1.3)), if

$$2^{(1/2)(s-2)} \Gamma(\tfrac{1}{2}s) \sqrt{\frac{2}{\pi}}\, \beta_1^{(1/2)(s-1)} \geq 1,$$

as required.

8.3. Proof of Part (i) of the Theorem

We first have to put the equation in an alternative equivalent form. Thus we write

$$\begin{aligned}
f(x) &= \sqrt{\frac{2}{\pi}} \int_0^\infty e^{-x^2/2(\beta_0+\beta_1 v^2)} \frac{f(v)\,dv}{\sqrt{\beta_0 + \beta_1 v^2}} \\
&= \sqrt{\frac{2}{\pi}} \int_0^\infty e^{-x^2/2\beta_1 v^2} \frac{f(v)\,dv}{\sqrt{\beta_0 + \beta_1 v^2}}
\end{aligned}$$

$$+\sqrt{\frac{2}{\pi}}\int_0^\infty\left(e^{-x^2/2(\beta_0+\beta_1 v^2)}-e^{-x^2/2\beta_1 v^2}\right)\frac{f(v)\,dv}{\sqrt{\beta_0+\beta_1 v^2}}$$
$$= T_1 f + T_2 f, \tag{8.3.1}$$

say, thus defining the linear operators T_1, T_2. Given s with $s > 1$, we investigate the properties of T_1, T_2 in the Banach space B_s of continuous functions f such that $f \in L(0,\infty)$ and also

$$f(x) = x^{-s}\,g(x), \qquad g \text{ bounded},$$

with the obvious norm

$$||f||_s = \sup|x^s f(x)| + \int_0^\infty |f|\,dx.$$

We define the norm of an operator in the usual way:

$$|||T|||_s = \sup_{f\in B_s}\left(||Tf||_s/||f||_s\right).$$

We first assert T_1 is bounded, indeed

$$|||T_1|||_s < 1, \tag{8.3.2}$$

if, given β_1, s is chosen so that (8.1.1) is satisfied, i.e.,

$$2^{(1/2)(s-2)}\Gamma(\tfrac{1}{2}s)\sqrt{\frac{2}{\pi}}\,\beta_1^{(1/2)(s-1)} < 1. \tag{8.3.3}$$

Indeed,

$$\begin{aligned}
|(T_1 f)(x)| &\le \sqrt{\frac{2}{\pi}}\int_0^\infty e^{-x^2/2\beta_1 v^2}\frac{v^{-s}|g(v)|\,dv}{\sqrt{\beta_0+\beta_1 v^2}}, \qquad f(v) = v^{-s}g(v),\\
&\le \sqrt{\frac{2}{\pi}}\int_0^\infty e^{-x^2/2\beta_1 v^2}\beta_1^{-1/2}v^{-s-1}|g(v)|\,dv\\
&\le (\sup|g|)\sqrt{\frac{2}{\pi}}\,\beta_1^{(1/2)s-1/2}\,x^{-s}\int_0^\infty e^{-(1/2)\tau^2}\tau^{s-1}\,d\tau, \quad \beta^{1/2}v = x/\tau,
\end{aligned}$$

so that

$$|x^s(T_1 f)(x)| \le \sup|g|2^{1/2(s-2)}\,\Gamma(\frac{1}{2}s)\sqrt{\frac{2}{\pi}}\beta_1^{1/2(s-1)}. \tag{8.3.4}$$

Also, essentially as we observed in the proof of part (ii) of the theorem,

$$\int_0^\infty |T_1 f|\,dx \le \int_0^\infty |f|\,dx, \tag{8.3.5}$$

and the combination of (8.3.4) and (8.3.5) proves (8.3.2).

We assert also that T_2 is compact in B_s, where we continue to assume (8.3.3) to be true. To do this, we have to show that, if $\{f_n\}$ is a bounded sequence in B_s, then we can find a subsequence of $\{T_2 f_n\}$ which is convergent.

We will establish the following facts. If $\{f_n\}$ is bounded in B_s, so that, for some $K > 0$,

$$|f_n| \leq K x^{-s}, \qquad \int_0^\infty |f_n| \leq K,$$

then there exist constants K^*, δ, dependent only on K, such that

$$|T_2 f_n| \leq K^* x^{-s-\delta} \qquad \text{for } x \geq 1, \tag{8.3.6}$$

$$|T_2 f_n| \leq K^* \qquad \text{for all } x, \tag{8.3.7}$$

$$\int_0^\infty |T_2 f_n| \leq K^*, \tag{8.3.8}$$

$$|(T_2 f_n)'| \leq K^*(x + x)^{-1} \qquad \text{for all } x. \tag{8.3.9}$$

From these it follows easily that $\{T_2 f_n\}$ is convergent in B_s. For, in any compact subset of $(0, \infty)$, $\{T_2 f_n\}$ is uniformly bounded and equicontinuous, and so, by Ascoli, there exists a subsequence (which without ambiguity we continue to denote by $\{T_2 f_n\}$) such that $\{T_2 f_n\}$ is uniformly convergent (in the usual sup norm) in the subset. Thus, given $\epsilon > 0$, we can choose x_0 sufficiently small and X sufficiently large that, from (8.3.6) and (8.3.7),

$$|T_2 f_n - T_2 f_m| < \tfrac{1}{2} s x^{-s} \qquad \text{for} \quad x \leq x_0, \quad x \geq X.$$

With x_0, X thus chosen, we can, from the uniform convergence in $[x_0, X]$, conclude that

$$||T_2 f_n - T_2 f_m||_s < \epsilon,$$

as required.

It remains therefore to prove (8.3.6)–(8.3.9). We remark first that (8.3.8) follows as in the proof of part (ii). Indeed, since the quantity $(\cdots)$ in the definition of T_2 is positive, we have

$$\int_0^\infty |T_2 f| \leq \int_0^\infty |f|.$$

To prove (8.3.7),

$$|T_2 f| \leq \sqrt{\frac{2}{\pi \beta_0}}\, \beta_0 \int_0^\infty |f(v)|\, dv \leq \sqrt{\frac{2}{\pi \beta_0}} K,$$

and differentiating the definition of T_2, we easily obtain (8.3.9). Finally, to prove (8.3.6) we have

$$
\begin{aligned}
|T_2 f| \leq & \sqrt{\frac{2}{\pi}} \int_0^{\beta_0+\beta_1 v^2 = x} e^{-x^2/2(\beta_0+\beta_1 v^2)} \frac{|f(v)\, dv}{\sqrt{\beta_0+\beta_1 v^2}} \\
& + \sqrt{\frac{2}{\pi}} \int_{\beta_0+\beta_1 v^2 = x}^{\infty} e^{-x^2/2\beta_1 v^2} \left\{ e^{-(x^2/2(\beta_0+\beta_1 v^2)) + (x^2/2\beta_1 v^2)} - 1 \right\} \frac{|f(v)|\, dv}{\sqrt{\beta_0+\beta_1 v^2}} \\
\leq & \sqrt{\frac{2}{\pi \beta_0}}\, e^{-x/2} \int_0^{\infty} |f(v)|\, dv + O\left\{ \int_{\beta_0+\beta_1 v^2 = x}^{\infty} e^{-x^2/2\beta_1 v^2} \left(\frac{x^2}{v^4}\right) \frac{v^{-s}|g(v)|}{v} \right\},
\end{aligned}
$$

since we are interested now only in x large,

$$
\leq \sqrt{\frac{2}{\pi \beta_0}} e^{-x/2} K + K^* x^2 \int_{\beta_0+\beta_1 v^2 = x}^{\infty} e^{-x^2/2\beta_1 v^2} v^{-s-5}\, dv, \qquad (8.3.10)
$$

for some choice of K^*, since g is bounded. If we make the change of variable $v = x/\tau$ in the last integral, we see that the second term on the right of (8.3.10) can be bounded by $K^* x^{-s-2}$, with a possibly changed value of K^*, from which (8.3.6) follows immediately.

We have thus expressed our equation in the form (8.3.1), where $|||T_1||| < 1$ and T_2 is compact, and both T_1 and T_2 are positive (in that they carry positive functions to positive functions). Thus we can write

$$
(I - T_1) f = T_2 f,
$$

or, since $|||T_1||| < 1$,

$$
f = (I - T_1)^{-1} T_2 f = T f, \quad \text{say,}
$$

and the operator on the right is necessarily compact, and also positive, as we see by expanding $(I - T)^{-1}$ in a binomial series. By [3, Theorem 2.5], we know that $(I - T_1)^{-1}\, T_2 = T$ has an eigenvalue with nonnegative eigenfunction provided that there is some strictly positive ϕ such that

$$
T^p \phi \geq \alpha \phi \qquad (8.3.11)
$$

for some integer p and some strictly positive number α. Once we have such an eigenfunction, we have a nonnegative (and so strictly positive) solution of (8.1.2) lying in B_s. (The eigenvalue has to be unity, seen by integrating both sides over $(0, \infty)$.)

To prove (8.3.11), we take

$$
\phi(x) = \begin{cases} 1, & x \leq 1, \\ x^{-s}, & x \geq 1. \end{cases}
$$

Using the sort of computations already employed, it is easy to see that (8.3.11) is satisfied, with $p = 1$.

We have thus established the existence of a solution in B_s for any s satisfying (8.1.1). Since it is in B_s, we can normalize it so that (8.1.3) holds.

We next want to show that the solution is unique, provided that (8.1.3) holds, and we need to establish first that any positive solution lies in B_s for any s such that (8.1.1) is true. We establish in the Appendix that, if $\beta_1 < 2e^\gamma$, where γ is Euler's constant, then there is a unique solution $s_0 > 1$ to the equation

$$(2\beta_1)^{(s_0-1)/2}\Gamma(\tfrac{1}{2}s_0) = \sqrt{\pi}, \tag{8.3.12}$$

and that the set of s (> 1) for which (given β_1) (8.1.1) is true is the set $1 < s < s_0$. (Note of course that (8.3.12) is satisfied by $s_0 = 1$.) We thus want to establish that any positive solution lies in B_s for any s with $1 < s < s_0$.

To do this, we form the Mellin transform of a solution,

$$\hat{f}(s) = \int_0^\infty x^{s-1} f(x)\,dx.$$

This certainly exists for $0 < R(s) \le 1$, and represents an analytic function of s for $0 < R(s) < 1$. We now transform equation (8.1.2), in the form (8.3.1). We have

$$\begin{aligned}
\int_0^\infty x^{s-1}(T_1 f)(x)\,dx &= \sqrt{\frac{2}{\pi}}\int_0^\infty \frac{f(v)\,dv}{\sqrt{\beta_0+\beta_1 v^2}}\int_0^\infty e^{-x^2/2\beta_1 v^2} x^{s-1}\,dx \\
&= \sqrt{\frac{2}{\pi}}\int_0^\infty \frac{\beta_1^{(1/2)s} v^s f(v)\,dv}{\sqrt{\beta_0+\beta_1 v^2}}\int_0^\infty y^{s-1}\,e^{-(1/2)y^2}\,dy \\
&= 2^{(1/2)(s-2)}\,\Gamma(\tfrac{1}{2}s)\sqrt{\frac{2}{\pi}}\,\beta_1^{(1/2)(s-1)}\int_0^\infty \frac{v^{s-1} f(v)\,dv}{\sqrt{1+\beta_0/(\beta_1 v^2)}}.
\end{aligned}$$

Since

$$\sqrt{1+\beta_0/(\beta_1 v^2)} = 1 + O(v^{-2})$$

for large v, we see that the last line can be written as

$$2^{(1/2)(s-2)}\Gamma(\tfrac{1}{2}s)\sqrt{\frac{2}{\pi}}\beta_1^{(1/2)(s-1)}\int_0^\infty v^{s-1} f(v)\,dv + F_1(s),$$

where F_1 is analytic for $R(s) < 3$. Similarly, the previous estimates for T_2 show that

$$\int_0^\infty x^{s-1}(T_2 f)(x)\,dx = F_2(s),$$

where $F_2(s)$ is analytic for $R(s) < 3$. Thus (8.3.1) transforms to

$$\int_0^\infty x^{s-1} f(x)\,dx\left\{1 - (2\beta)_1^{(1/2)(s-1)}\Gamma(\tfrac{1}{2}s)/\sqrt{\pi}\right\} = F_1(s) + F_2(s), \tag{8.3.13}$$

and so $\hat{f}(s)$ can be continued analytically to $R(s) < 3$ except for possible singularities at the zeros of

$$(2\beta_1)^{(1/2)(s-1)}\,\Gamma(\tfrac{1}{2}s) = \sqrt{\pi}. \tag{8.3.14}$$

There is a zero at $s = 1$, but $F_1(s) + F_2(s) = 0$ at $s = 1$. (This is a consequence of (8.3.13) and the fact that $f \in L_1$.) Hence $\hat{f}(s)$ has no singularity at $s = 1$, and the next zero of (8.3.14), as we increase $R(s)$, is at $s = s_0$. (The discussion of F_1 and F_2 shows that they are always analytic in a wider range than $\hat{f}$, so that the first singularity of $\hat{f}$ occurs at the first zero of (8.3.14). Further, the first singularity of $\hat{f}$, since $f \geq 0$, must occur at s real, and so at s_0.) Hence $\hat{f}(s)$ can be continued analytically some way into $R(s) > s_0$ except for a pole at $s = s_0$, and from the inverse theorem for the Mellin transform we then have, as $x \to \infty$,

$$f(x) \sim Ax^{-s_0}.$$

We further note that $A > 0$. For $A = 0$ would imply that $\hat{f}(s)$ has no singularity at $s = s_0$, and so $f(x) = O(x^{-s})$ for some $s > s_0$, which is excluded by part (ii).

Having established asymptotic behavior, we can now prove that the solution is unique. Suppose for contradiction that we have two solutions f_1, f_2 with

$$f_1(x) \sim A_1 x^{-s_0}, \qquad f_2(x) \sim A_2 x^{-s_0}.$$

Since a change in A_1 merely multiplies f_1 by a multiplicative constant, we can choose A_1 sufficiently large that $f_1 > f_2$ everywhere. Now reduce A_1 until $f_1 - f_2$ is nonnegative with a zero minimum. If this minimum occurs at a finite point x_0, then since $f_1 - f_2$ satisfies (8.1.2), we have a contradiction when the equation is evaluated at x_0. If the minimum occurs at infinity, then $f_1 - f_2 \geq 0$, $f_1 - f_2 = o(x^{-s_0})$ as $x \to \infty$. As remarked above, this is impossible. This concludes the proof of uniqueness and so of the theorem.

8.4. Appendix

We show that the equation

$$(2\beta_1)^{(s_0-1)/2}\,\Gamma(\tfrac{1}{2}s_0) = \sqrt{\pi} \tag{A1}$$

has a unique solution with $s_0 > 1$ if

$$\beta_1 < 2e^{\gamma},$$

and no such solution if $\beta_1 \geq 2e^{\gamma}$.

We can write the left-hand side of (A1) (for general s) as

$$\begin{aligned}(2\beta_1)^{(s-1)/2}\int_0^\infty t^{(1/2)s-1}e^{-t}\,dt &= (2\beta_1)^{1/2}\int_0^\infty (2\beta_1 t)^{(1/2)s-1}e^{-t}\,dt \\ &= (2\beta_1)^{-1/2}\int_0^\infty u^{(1/2)s-1}e^{-u/(2\beta_1)}\,du \\ &= F(s),\end{aligned}$$

say. Since each differentiation with respect to s introduces a factor $\log\, u$ in the integrand, we see that $F''(s) > 0$. Also, $F(1) = \sqrt{\pi}$, and so, if $F'(1) \geq 0$, then $F(s) > \sqrt{\pi}$ for $s > 1$, and we have no solution to (A1) for $s > 1$. On the other hand, if $F'(1) < 0$, then, since $F(s) \to \infty$ as $s \to \infty$, we must have one (and only one) solution of (A1) for $s > 1$. The critical value of β is thus the value for which $F'(1) = 0$, i.e., for which

$$\tfrac{1}{2}\log\,(2\beta_1)\Gamma(\tfrac{1}{2}) + \tfrac{1}{2}\Gamma'(\tfrac{1}{2}) = 0,$$

i.e.,

$$\log\,(2\beta_1) = -\Gamma'(\tfrac{1}{2})/\Gamma(\tfrac{1}{2}).$$

Now from Legendre's duplication formula,

$$\Gamma(z)\Gamma(z+\tfrac{1}{2})2^{2z-1} = \sqrt{\pi}\,\Gamma(2z),$$

we can differentiate and set $z = \frac{1}{2}$, so that, since $\Gamma(1) = 1$, $\Gamma(\frac{1}{2}) = \sqrt{\pi}$,

$$\Gamma'(\tfrac{1}{2}) = \sqrt{\pi}(\Gamma'(1) - 2\log\,2).$$

Also, from

$$\frac{1}{\Gamma(z)} = ze^{\gamma z}\prod_{r=1}^{\infty}\left\{\left(1+\frac{z}{r}\right)e^{-z/r}\right\},$$

we have

$$\frac{\Gamma'(1)}{\Gamma(1)} = -\gamma, \quad \Gamma'(1) = -\gamma.$$

Thus the critical value of β_1 is

$$\log(2\beta_1) = 2\log\,2 + \gamma,$$
$$\beta_1 = 2e^{\gamma}.$$

References

[1] Engle, R.F. Autoregressive conditional heteroscedasticity with estimates of United Kingdom inflation. *Econometrica*, **50** (1982), 987–1007.

[2] Pantula, S.G. Estimation of autoregressive models with ARCH errors. *Sankhya B*, **50** (1988), 119–138.

[3] Krasnosel'skii, M.A. *Positive Solutions of Operator Equations*. Noordhoff, 1964.

9
Self-Similar Solutions of the Second Kind

L.A. Peletier

9.1. Introduction

In recent years there has been a surge of interest in self-similar solutions of the *second kind*. Such solutions are not newly discovered; they had been identified and in fact so named by Zel'dovich in 1956, in the context of a variety of problems, such as shock waves in gas dynamics [18], [34], [35], and filtration through elastoplastic materials [5], [10], [7, 8]. We cite from Zel'dovich's Foreword to Barenblatt's book on intermediate asymptotics [5]:

> We shall reserve the name *solutions of the second kind* for the large and ever-growing class of solutions for which the exponents are found in the process of solving the problem, analogous to the determination of eigenvalues for linear equations. For this case, conservation laws and dimensional considerations prove to be insufficient.

What makes these solutions interesting at present is that they lead to nonlinear eigenvalue problems, which, with the knowledge acquired over the last decades, have become much more accessible and tractable than in the 1950s.

In this chapter we recall some self-similar solutions of the first and second kind, citing a few classical examples and a few more recent ones, and then sketch the progress which has recently been made on some of these problems.

As a first example of a self-similar solution of the *first kind* we mention the fundamental solution of the Heat Equation. Thus, consider the Cauchy Problem

$$\begin{cases} u_t = \Delta u & \text{in } \mathbf{R}^N \times \mathbf{R}^+, \\ u(x,0) = u_0(x) & \text{in } \mathbf{R}^N, \end{cases}$$

in which u_0 is concentrated at the origin so that $u_0(x) = 0$ if $x \neq 0$. Scale invariance suggests separation of variables:

$$u(x,t) = t^{-\alpha/2} f(\eta), \qquad \eta = |x|t^{-\beta}, \tag{9.1.1}$$

where α and β are parameters which need to be determined.

(a) For the *Ansatz* (9.1.1) to yield an equation involving only f and η we need to set

$$\beta = \tfrac{1}{2}.$$

(b) If we require *mass conservation*, we need to choose α and β such that

$$\frac{\alpha}{2} = N\beta.$$

With these values for α and β we obtain the solution

$$E(x,t) = t^{-N/2} f(\eta), \qquad \eta = |x| t^{-1/2},$$

in which

$$f(\eta) = Ae^{-\eta^2/4}$$

and the constant A is adjusted according to the initial mass.

As a second example we consider the same problem but for the Heat Equation with absorption:

$$u_t = \Delta u - |u|^{p-1}u \qquad \text{in} \quad \mathbf{R}^N \times \mathbf{R}^+. \tag{9.1.2}$$

With the same Ansatz we find that we must choose

$$\alpha = \frac{2}{p-1}, \qquad \beta = \tfrac{1}{2},$$

so that the parabolic equation for u is reduced to an elliptic one for f.

In these examples α and β are both determined from simple scaling rules. This characterizes the self-similar solutions of the first kind.

The situation is different for the Barenblatt Equation for elastoplastic filtration [10], [11]. This equation describes the flow of an elastic fluid through an elastoplastic porous medium. Typical of such a medium is that the compressibility depends on whether the pressure of the fluid increases or decreases. Thus we obtain the equation

$$u_t = \begin{cases} \kappa_+ \Delta u & \text{if } u_t > 0, \\ \kappa_- \Delta u & \text{if } u_t < 0, \end{cases}$$

for the pressure u, in which κ_+ and κ_- are positive constants. With a suitable scaling of the time, we can write this equation in the form

$$u_t + \gamma |u_t| = \Delta u \quad \text{in}\, \mathbf{R}^N \times \mathbf{R}^+, \qquad -1 < \gamma < 1, \tag{9.1.3}$$

in which $\gamma = (\kappa_- - \kappa_+)/(\kappa_- + \kappa_+)$.

Substitution of the Ansatz (9.1.1) leads again to the requirement that $\beta = \frac{1}{2}$. However, there is no mass conservation to determine α. This can be seen from the following simple formal argument. Integrating equation (9.1.3) over $\mathbf{R}^N$ and changing the order of differentiation and integration we obtain

$$\frac{d}{dt}\int u = -\gamma \int |u_t|,$$

so that mass is only conserved if $\gamma = 0$.

Finding the value of the exponent α in this problem requires a detailed analysis of the nonlinear ordinary differential equation we obtain for f in which α appears as an eigenvalue. Self-similar solutions for which the exponent needs to be so determined are called of the *second kind* and the exponent involved is called *anomalous*.

A second, equally classical, example arises in the description of the spread of an instantaneous planar source of turbulence in a quiescent fluid [6], [9]. This problem leads to the nonlinear diffusion equation

$$\frac{\partial q}{\partial t} = \frac{\partial}{\partial z}\left(K_q \frac{\partial q}{\partial z}\right) - \varepsilon_t, \qquad z \in \mathbf{R}, \qquad t > 0, \tag{9.1.4}$$

in which q denotes the turbulent energy density, K_q the turbulent energy diffusion coefficient, and ε_t the rate of energy dissipation. It is assumed that initially the turbulence was concentrated in a layer centered on the plane $z = 0$. On the basis of Kolmogorov's similarity hypothesis, and dimensional analysis we are led to the following expressions for K_q and ε_t:

$$K_q = \ell(t)\sqrt{q} \qquad \text{and} \qquad \varepsilon_t = c\frac{q^{3/2}}{\ell(t)}, \tag{9.1.5}$$

where c is a positive universal constant and $\ell(t)$ denotes a characteristic length. Since this equation is degenerate its solution will have compact support at any time $t > 0$ if its support is compact initially. This suggests the following choice for $\ell(t)$:

$$\ell(t) = \alpha h(t), \qquad \text{where} \quad h(t) = \tfrac{1}{2}\text{meas}\{\text{supp}\, q(z,t)\}, \tag{9.1.6}$$

in which α is another constant which needs to be determined from a more detailed theory. Under some mild restrictions on the initial value $q(z, 0)$ [24] the solution of the Cauchy Problem for (9.1.4) converges to a self-similar solution of the form

$$q(z,t) = A^2 t^{-2\mu} f(\eta), \quad \eta = \frac{z}{h(t)}, \qquad h(t) = At^{1-\mu}, \quad \mu > 0. \tag{9.1.7}$$

This self-similar solution is again of the second kind, the exponent μ having to be determined from the nonlinear eigenvalue problem for $f(\eta)$.

It is interesting to note here the recent work on the related $k - \varepsilon$ model for the spead of turbulent bursts which also leads to self-similar solutions of the second kind [12], [20], [26].

We mention two further recent examples of self-similar solutions of the second kind, both involving the Porous Media Equation [1], [27]

$$u_t = \Delta(u^m), \qquad m > 0. \tag{9.1.8}$$

To study the regularity of solutions in domains with several space dimensions when $m > 1$, Aronson and Graveleau [2] constructed a family of radially symmetric self-similar solutions which initially vanish identically in a ball around the origin. As time progresses this ball, in which the solution remains zero, contracts until after a finite time T it becomes a point. Their solution is of the form

$$v(x,t) = \frac{r^2}{T-t}\varphi(\eta) \qquad \text{and} \qquad \eta = -(T-t)|x|^{-\alpha}, \tag{9.1.9}$$

where $v = [m/(m-1)]u^{m-1}$ denotes the pressure. This self-similar solution is of the second kind and α is the anomalous exponent, which, they show, satisfies

$$1 + \frac{(m-1)(N-2)}{(m-1)(N+2)+4} < \alpha < 2.$$

For another discussion of this problem we refer to [14].

It is well known that solutions of the Cauchy problem for (9.1.8) die out in finite time when $N \geq 3$ and $m < 1 - 2/N$. In order to investigate the behavior of the solution near the extinction time T, King [25] and Philip [31] proposed a family of radially symmetric self-similar solutions of the form

$$u(x,t) = (T-t)^{\alpha} f(\eta), \qquad \eta = |x|(T-t)^{-\beta}, \tag{9.1.10}$$

in which $\alpha = (1-2\beta)/(1-m)$. Such solutions are also of the second kind. They were recently analyzed by Mark Peletier and Hongfei Zhang [30] who proved that for each $N \geq 3$ and each $0 < m < 1 - 2/N$ there exists a unique pair (α, β) for which (9.1.10) yields a solution of (9.1.8), and confirmed King's conjecture that

$$0 < \alpha < \frac{N-2}{(1-m)N-2}.$$

These solutions were shown to be the asymptotic profiles near the extinction time [15].

In the following sections we will sketch the analysis involved in the study of the self-similar solutions of the Barenblatt Equation and of the turbulence problem.

9.2. The Barenblatt Equation

When we substitute the Ansatz (9.1.1) into (9.1.3) we find that the function f needs to satisfy the following two-point boundary value problem

$$\text{(I)} \quad \begin{cases} f'' + \frac{N-1}{\eta} f' = \frac{1}{2} F\big(-(\eta f' + \alpha f)\big), \quad f > 0, \quad 0 < \eta < \infty, & (9.2.1) \\ f(0) = 1, \quad f'(0) = 0, & (9.2.2) \\ \eta^{\alpha} f(\eta) \to 0 \quad \text{as}\ \eta \to \infty, & (9.2.3) \end{cases}$$

in which

$$F(s) = s + \gamma|s|,$$

and the condition at infinity follows from the initial condition

$$\lim_{t\downarrow 0} u(x,t) = |x|^{-\alpha} \lim_{\eta\to\infty} \eta^\alpha f(\eta) = 0 \qquad \text{when} \qquad x \neq 0.$$

Because equation (9.1.1) is invariant under multiplication, the initial condition $f(0) = 1$ involves no loss of generality.

In [13] it was shown that for every $\gamma \in (-1, 1)$, there exists a unique number $\alpha(\gamma) \in ((N-2)_+, \infty)$ for which problem (I) has one and only one solution $f(\eta)$. The asymptotic behavior of $f(\eta)$ was shown to be

$$f(\eta) = A\eta^{\alpha-N} \exp\left(-\frac{1+\gamma}{4}\eta^2\right) \cdot \{1 + O(\eta^{-2})\} \qquad \text{as} \quad \eta \to \infty$$

for some positive constant A.

Thus the function $\alpha(\gamma)$ is well defined on $(-1, 1)$. It was shown to have the following properties:

(1) $\alpha(\gamma)$ is analytic for $-1 < \gamma < 1$ [3, 4].

(2) If $\gamma_1 < \gamma_2$, then $\alpha(\gamma_1) < \alpha(\gamma_2)$.

(3) $$\alpha(\gamma) \to \begin{cases} N-2 & \text{if} \quad N > 2, \\ 0 & \text{if} \quad N = 2, \end{cases} \qquad \text{as } \gamma \to -1.$$

When $\gamma = 0$, then (9.1.3) reduces to the Heat Equation, so that

(4a) $$\alpha(0) = N.$$

A more detailed study of the behavior of $\alpha(\gamma)$ near $\gamma = 0$ was made in [17] for $N = 1$ and subsequently by [3, 4] for $N \geq 2$. It was found that

(4b) $$\left.\frac{d\alpha}{d\gamma}\right|_{\gamma=0} = \frac{2(N/2)^{N/2}}{e^{N/2}\Gamma(N/2)}.$$

At the other end, as $\gamma \to 1$, then $\alpha(\gamma) \to \infty$. More precisely [28],

(5) $$\alpha(\gamma) \sim \frac{\rho_\nu}{\sqrt{2}} \frac{1}{\sqrt{1-\gamma}} \qquad \text{as} \quad \gamma \to 1,$$

where ρ_ν is the first zero of the Bessel function J_ν and $\nu = N/2 - 1$.

The graph of $\alpha(\gamma)$ is sketched in Fig. 9.1.

The analysis of the eigenvalue problem (9.2.1–9.3) is easiest carried out after transforming (9.2.1) to a Riccati equation by introducing the variables

$$t = \frac{\eta^2}{4} \qquad \text{and} \qquad y(t) = -\frac{\eta f'}{f}. \tag{9.2.4}$$

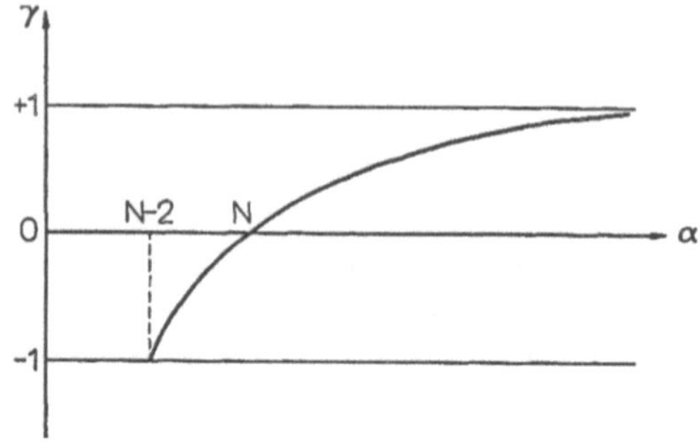

Fig. 9.1 The graph of $\alpha(\gamma)$.

This yields the problem

$$\begin{cases} y' = -F(y-\alpha) + \frac{y}{2t}\{y-(N-2)\}, & (9.2.5) \\ y(0) = 0 \quad \text{and} \quad y(t) \sim 2(1+\gamma)t \quad \text{as} \quad t \to \infty. & (9.2.6) \end{cases}$$

Following the orbit from the origin, it has to cross the line $\{y = \alpha\}$, say, for the first time at $t = \tau$. Then

$$y'(\tau) = \frac{\alpha}{2\tau}\{\alpha - (N-2)\} > 0,$$

where we have inequality because of uniqueness. Therefore a necessary condition on α is

$$\alpha > N - 2.$$

The key ingredients used in proving that there exists only one value of α for every $\gamma \in (-1, 1)$ and the monotonicity of α with respect to γ, are the observations that the right-hand side of (9.2.5) is strictly increasing with respect to α and strictly decreasing with respect to γ.

The behavior near $\gamma = 0$, or $\alpha = N$, was proved by Goldenfeld et al. by means of renormalization group arguments and subsequently by Aronson and Vazquez who used arguments involving the Implicit Function Theorem.

Finally, the asymptotic behavior of $\alpha(\gamma)$ as $\gamma \to 1$ was proved by a careful study of (9.2.5–9.2.6) involving a sequence of different scaling arguments.

9.3. Decay of Turbulent Bursts

When we apply the Ansatz (9.1.7) to (9.1.4), in which (9.1.5) and (9.1.6) have been used to eliminate K_q and ε_t, we arrive at the following two-point boundary value problem for the function $(4\alpha^2/9)f(\eta)$, which we denote again by f:

$$\begin{cases} (f^{3/2})'' + (1-\mu)\eta f' + 2\mu f - \lambda f^{3/2} = 0, & 0 < \eta < 1, \quad (9.3.1) \\ f > 0, & 0 \le \eta < 1, \quad (9.3.2) \\ (f^{3/2})'(0) = 0 \quad \text{and} \quad f(1) = 0, \quad (f^{3/2})'(1) = 0. & \quad (9.3.3) \end{cases}$$

Here we have combined the coefficients c and α into a single eigenvalue parameter

$$\lambda = \frac{3c}{2\alpha^2}.$$

Remark. It is interesting to observe that the *Very Singular Solution* of the nonlinear diffusion equation with absorption

$$u_t = (u^m)_{xx} - u^p, \qquad u \ge 0,$$

which is a self-similar solution (of the first kind) of the form,

$$u(x,t) = t^{1/(p-1)} f(\eta), \qquad \eta = x/t^{2(p-1)/(p-m)},$$

leads to an analogous problem for f [29], which has a solution if and only if $m < p < m+2$. The additional freedom afforded by the function $h(t)$ in (9.1.4) and (9.1.5) allows problem (9.3.1)–(9.3.3) to have a solution although here we have $p = m = \frac{3}{2}$.

Problem (9.3.1)–(9.3.3) has two eigenvalues, λ and μ, where μ plays the role of the anomalous exponent. The principal question is to find for which values of λ and μ this problem possesses a solution. This suggests we introduce the set

$$C = \{(\lambda, \mu) : \textit{there exists a solution of (9.3.1)–(9.3.3)}\}.$$

Some preliminary observations can readily be made.

If $\lambda = 0$, then problem (9.3.1)–(9.3.3) can be solved explicitly when we set $\mu = \frac{1}{3}$. The resulting solution is

$$f(\eta) = \frac{1}{81}(1-\eta^2)^2. \qquad (9.3.4)$$

Therefore the point $(\lambda, \mu) = (0, \frac{1}{3})$ belongs to C.

Inspecting the differential equation (9.3.1) we find that if $\lambda > 0$, then any solution f must have the properties

$$f(\eta) < \left(\frac{2\mu}{\lambda}\right)^2 \quad \text{and} \quad f'(\eta) < 0 \quad \text{for} \quad 0 < \eta < 1. \qquad (9.3.5)$$

When we integrate the differential equation over the interval $(0, 1)$ we obtain

$$(1-\mu)\int_0^1 \eta f' = \int_0^1 f(\lambda\sqrt{f} - 2\mu). \qquad (9.3.6)$$

Since by (9.3.5) the first integral is negative and the second one is positive it follows that $\mu < 1$ for all $\lambda > 0$.

Integrating the left-hand side of (9.3.6) by parts we obtain

$$(3\mu - 1)\int_0^1 f = \lambda \int_0^1 f^{3/2}. \tag{9.3.7}$$

Because both integrals are positive this implies that $\mu > \frac{1}{3}$ for all $\lambda > 0$ and $\mu = \frac{1}{3}$ for $\lambda = 0$.

Summarizing we conclude that $C \subset \{\lambda > 0,\ \frac{1}{3} < \mu < 1\} \cup \{(0, \frac{1}{3})\}$.

A detailed analysis [19] shows that:

(1) For each $\lambda \geq 0$ there exists a unique $\mu(\lambda) \in (\frac{1}{3}, 1)$ for which problem (9.3.1)–(9.1.3) has a solution.

(2) $\mu(\lambda)$ is continuous for $0 \leq \lambda < \infty$.

(3) If $\lambda_1 < \lambda_2$, then $\mu(\lambda_1) < \mu(\lambda_2)$.

(4) $$\frac{3\sqrt{3}+\lambda}{9\sqrt{3}+2\lambda} < \mu(\lambda) < \frac{3+\lambda}{9+\lambda} \quad \text{for} \quad \lambda > 0.$$

(5) $$\frac{4}{7+\sqrt{1+4\lambda}} < 1 - \mu(\lambda) < \frac{2\sqrt{6}}{\sqrt{\lambda}} \quad \text{for} \quad \lambda > 0.$$

(6) $$\mu(\lambda) = 1 - \frac{2}{\sqrt{\lambda}}[1 + o(1)] \quad \text{as} \quad \lambda \to \infty.$$

Note that (6) implies that the lower bound in (5) is sharp.

Finally, in [13] it is shown that

(7) $$\mu(\lambda) = \frac{1}{3} + \frac{2}{63}\lambda + \mathcal{O}(\lambda^2) \quad \text{as} \quad \lambda \to 0.$$

These results are sketched in Fig. 9.2.

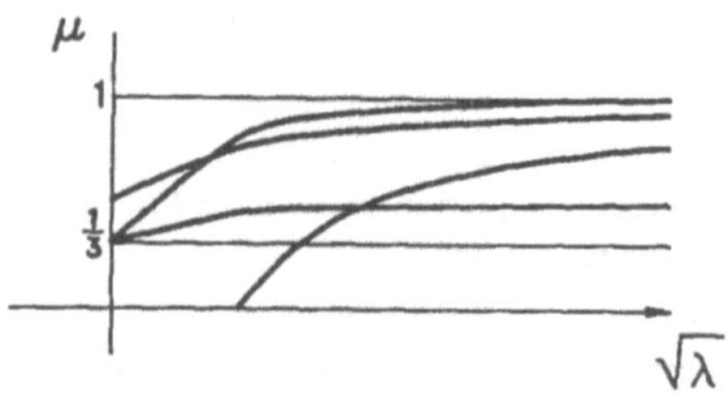

Fig. 9.2. The solution set C.

The main ingredients in the proofs of (1)–(6) are a shooting argument (existence) and convexity arguments (uniqueness), applied to the equation we obtain after transformation to the function $g(\eta) = \frac{1}{2}\sqrt{f(\eta)}$:

$$g'' + 2\frac{g'}{g}\left(g' + \frac{1-\mu}{6}\eta\right) + \frac{\mu - \lambda g}{3} = 0, \qquad 0 < \eta < 1. \tag{9.3.8}$$

They are based on the observation that if for a given value of μ there exist two eigenvalues λ_1 and λ_2 such that (9.3.8) has solutions g_1 and g_2 which satisfy the appropriate boundary conditions

$$g'(0) = 0 \qquad \text{and} \qquad g(1) = 0, \tag{9.3.9}$$

then

$$\lambda_1 < \lambda_2 \quad \Rightarrow \quad g_1''(\eta) < g_2''(\eta) < 0 \qquad \text{for } 0 < \eta < 1.$$

Because

$$g_1'(1) = g_2'(1) = -\frac{1-\mu}{6},$$

this is incompatible with the boundary condition (9.3.9) at $\eta = 0$. For the uniqueness of μ when λ is kept fixed, a similar argument applies.

The properties of the eigenvalue branch $C = \{(\lambda, \mu(\lambda)) : \lambda \geq 0\}$, and especially the bounds, follow from convexity arguments (lower bound) and from an integral identity obtained by multiplying (9.3.8) by $\eta^\beta g^2$, where $\beta = \frac{2}{1-\mu} - 3$, and integrating the result over (0, 1). Finally the asymptotic behavior of $\mu(\lambda)$ for large values of λ is obtained by suitably scaling the equation near the endpoint, where a boundary layer develops.

The local behavior near $\lambda = 0$ given in (7) is proved by renormalization group arguments, combined with perturbation theory. A further analysis of this problem has recently appeared in [32, 33].

References

[1] Aronson, D.G. The porous media equation. In *Nonlinear Diffusion Problems*. Lecture Notes in Mathematics, Vol. **1224**. Springer-Verlag, New York, 1986.

[2] Aronson, D.G. and J. Graveleau. A self-similar solution to the focusing problem for the porous media equation. *European J. Appl. Math.*, **4** (1993), 65–81.

[3] Aronson, D.G. and J.L. Vazquez. Calculation of anomalous exponents in nonlinear diffusion. *Phys. Rev. Lett.*, **72** (1994), 348–351.

[4] Aronson, D.G. and J.L. Vazquez. Anomalous exponents in nonlinear diffusion. *J. Nonlinear Sci.*, **5** (1995), 29–56.

[5] Barenblatt, G.I. Similarity, Self-Similarity, and Intermediate Asymptotics. Consultants Bureau, New York, 1979.

[6] Barenblatt, G.I. Self-similar turbulence propagation from an instantaneous plane source. In *Nonlinear Dynamics and Turbulence* (G.I. Barenblatt, G. Iooss, and D.D. Joseph, eds.), pp. 48–60. Pitman, London, 1983.

[7] Barenblatt, G.I., V.M. Entov, and V.M. Ryzhik. *Theory of Unsteady Filtration of Fluids and Gases* (in Russian). Nedra, Moscow, 1972.

[8] Barenblatt, G.I., V.M. Entov, and V.M. Ryzhik. *Theory of Fluid Flows Through Natural Rocks*. Kluwer Academic, Dordrecht, 1990.

[9] Barenblatt, G.I., N.L. Galerkina, and M.V. Luneva. Evolution of turbulent bursts. *Inzh.-Fiz. Zh. (J. Engrg. Phys.)*, **53** (1987), 733–740.

[10] Barenblatt, G.I. and A.P. Krylov. On elasto-plastic regime of filtration. *Izv. Akad. Nauk SSSR, OTN*, **2** (1955), 14–26.

[11] Barenblatt, G.I. and G.I. Sivashinsky. Self-similar solutions of the second kind in nonlinear filtration. *Appl. Math. Mech. PMM*, **33** (1969), 836–845.

[12] Bertsch, M., R. DalPasso, and R. Kersner. The evolution of turbulent bursts: The $b-\varepsilon$ model. *European J. Appl. Math.*, **5** (1994), 537–557.

[13] Chen, Lin-Yuan and N. Goldenfeld. Renormalisation group theory for the propagation of a turbulent burst. *Phys. Rev. A*, **45**, 5572.

[14] Diez, J.A., J. Gratton, and F. Minotti. Self-similar solutions of the second kind of nonlinear diffusion equations. *Quart. Appl. Math.*, **L** (1992), 401–414.

[15] Galaktionov, V.A. and L.A. Peletier. Asymptotic behavior near finite time extinction for the fast diffusion equation in $\mathbf{R}^N$. To appear in *Arch. Rational Mech. Anal.*

[16] Goldenfeld, N. *Lectures on Phase Transitions and the Renormalization Group*. Frontiers in Physics, vol. **85**. Addison-Wesley, Reading, MA, 1992.

[17] Goldenfeld, N., O. Martin, Y. Oono, and F. Liu. Anomalous dimensions and the renormalization group in a nonlinear diffusion. *Phys. Rev. Lett.*, **64** (1990), 1361–1364.

[18] Guderley, G. Starke Kugelige und Zylindrische Verdichtungsstösse in der Nähe des Kugelmittelpunktes bzw. der Zylinderachse. *Luftfahrtforschung*, **19** (1942), 302–312.

[19] Hastings, S.P. and L.A. Peletier. On self-similar solutions for the decay of turbulent bursts. *Eur. J. Appl. Math.*, **3** (1992), 319–341.

[20] Hulshof, J. Self-similar solutions of the $k-\varepsilon$ model for turbulence. *SIAM J. Math. Anal.* **28** (1997), 33–48.

[21] Kamenomostkaya, S.L. (S. Kamin). On a problem in the theory of filtration. *Dokl. Akad. Nauk USSR*, **116** (1957), 18–20.

[22] ———. Equation of elastoplastic mode of filtration. *Appl. Math. Mech. PMM*, **33** (1969), 1042–1049.

[23] Kamin, S., L.A. Peletier, and J.L. Vazquez. On the Barenblatt equation of elasto-plastic filtration. *Indiana Univ. Math. J.*, **40** (1991), 1333–1362.

[24] Kamin, S. and J.L. Vazquez. On the propagation of turbulent bursts. *European J. Appl. Math.*, **3** (1992), 263–272.

[25] King, J.R. Self-similar behavior for the equation of fast nonlinear diffusion. *Phil. Trans. Roy. Soc. London Ser. A*, **343** (1993), 337–375. Medskip.

[26] McLeod, J.B. Private communication.

[27] Peletier, L.A. The porous media equation. In *Applications of Nonlinear Analysis in the Physical Sciences* (H. Amann, N. Bazley, and K. Kirchgässner, eds.), pp. 229-241. Pitman, Boston, 1981.

[28] Peletier, L.A. The anomalous exponent of the Barenblatt equation. *European J. Appl. Math.*, **5** (1994), 165–175.

[29] Peletier, L.A. and D. Terman. *The very singular solution of porous media equation with absorption. J. Differential Equations*, **65** (1986), 396–410.

[30] Peletier, M.A. and Hongfei Zhang. Self-similar solutions of the fast diffusion equation which do not conserve mass. *Diff. Int. Eqns.*, **8** (1995), 2045–2064.

[31] Philip, J.R. Fast diffusion with loss at infinity. *J. Austral. Math. Soc. Ser. B*. To appear.

[32] Quiros, F. and J.L. Vazquez. Different similarity regimes for the turbulent burst equation. *C.R. Acad. Sci. Paris*, **320** (1995), 559–565.

[33] Quiros, F. and J.L. Vazquez. Self-similar turbulent bursts existence and analytic dependence. *Diff. Int. Equ.*, **8** (1995), 1677–1708.
[34] Zel'dovich, Ya. B. The motion of a gas under the action of a short time pressure (shock). *Akust. Zh.* **2** (1956), 28–38.
[35] Zel'dovich, Ya. B. and Yu. P. Raizer. Physics of shock waves and high-temperature hydrodynamic phenomena. Vol. II. Academic Press, New York, 1966.

10
On the Problem of a Moving Contact Angle

V.A. Solonnikov

10.1. Introduction

This chapter is a continuation of paper [1] devoted to the investigation of the asymptotics of a solution of the two-dimensional Navier–Stokes equations in the neighborhood of the point of a contact of the free boundary with a uniformly moving rigid wall. The contact angle is supposed to equal π. In [1] it was shown in particular that in the neighborhood of the contact point $x = 0$ the free surface is given by the equation of the type $h(x_1) = cx_1^{2-\kappa} + o(x_1^{2-\kappa})$, $x_1 \in (0, d)$, where c and κ are certain numbers depending on the velocity of the bottom $x_2 = 0$. In the present chapter, we estimate the constant c and obtain the information on its sign. This information can be used for the proof of the solvability of free boundary problems where such a contact occurs. In this chapter, we restrict ourselves to the simplest problem of filling an infinite plane capillary, without taking into consideration the force of gravity, and to the piston problem.

Suppose that a viscous incompressible capillary liquid occupies a region $\Omega \subset \mathbb{R}^2$ bounded by a straight line $\Sigma_d = \{x_2 = 0,\ -d < x_1 < 0\}$, a curve $\Gamma_d = \{x_2 = h(x_1),\ 0 < x_1 < d\}$, and by a circle $|\ x\ | = d$. The velocity vector field $\mathbf{v} = (v_1, v_2)$ and the pressure $p(x)$ satisfy the Navier–Stokes equations

$$-\nu\nabla^2\mathbf{v} + (\mathbf{v}\cdot\nabla)\mathbf{v} + \nabla p = 0, \qquad \nabla\cdot\mathbf{v} = 0, \qquad x \in \Omega_d, \quad (10.1.1)$$

and the boundary conditions

$$\mathbf{v}|_{\Sigma_d} = -\mathbf{a}, \qquad \mathbf{v}\cdot\mathbf{n}|_{\Gamma_d} = 0, \qquad \sigma H\mathbf{n} - T(\mathbf{v}, p)\mathbf{n}|_{\Gamma_d} = 0, \quad (10.1.2)$$

where $\mathbf{a} = (a, 0)$, $a > 0$, $\mathbf{n}$ and τ are the unit normal and tangential vector to Γ_d, respectively, i.e.,

$$\mathbf{n} = \left(-\frac{h'}{\sqrt{1+h'^2}}, \frac{1}{\sqrt{1+h'^2}}\right), \qquad \tau = \left(\frac{1}{\sqrt{1+h'^2}}, \frac{h'}{\sqrt{1+h'^2}}\right), \quad (10.1.3)$$

H is the curvature of Γ_d given by the formula

$$H = -\frac{d}{dx_1}\frac{h'}{\sqrt{1+h'^2}} = -\frac{h''}{(1+h'^2)^{3/2}},$$

and T is the stress tensor with the elements

$$T_{ij} = -p\delta_{ij} + \nu S_{ij}(\mathbf{v}), \qquad S_{ij} = \frac{\partial v_i}{\partial x_j} + \frac{\partial v_j}{\partial x_i}.$$

Finally, ν and σ are positive constant coefficients of viscosity and of the surface tension, respectively.

For a new vector field $\mathbf{w} = \mathbf{v} + \mathbf{a}$ relations (10.1.1), (10.1.2) assume the form

$$\begin{aligned} &-\nu\nabla^2\mathbf{w} - (\mathbf{a}\cdot\nabla)\mathbf{w} + (\mathbf{w}\cdot\nabla)\mathbf{w} + \nabla p = 0, \qquad \nabla\cdot\mathbf{w} = 0, \\ &\mathbf{w}|_{\Sigma_d} = 0, \qquad \mathbf{w}\cdot\mathbf{n} - an_1|_{\Gamma_d} = 0, \qquad \sigma H\mathbf{n} - T(\mathbf{w}, p)\mathbf{n}|_{\Gamma_d} = 0. \end{aligned} \tag{10.1.4}$$

Furthermore, if we differentiate the condition $\mathbf{w}\cdot\mathbf{n} - an_1|_{\Gamma_d} = 0$ along Γ_d, we obtain

$$\frac{\partial \mathbf{w}\cdot\mathbf{n}}{\partial\tau} = a\frac{\partial n_1}{\partial\tau} = \frac{a}{\sqrt{1+h'^2}}\frac{dn_1}{dx_1} = \frac{a}{\sigma}n_2\mathbf{n}\cdot T(\mathbf{w}, p)\mathbf{n}|_{x_2=h(x_1)}. \tag{10.1.5}$$

Let us map Ω_d onto a subdomain ω_d of the half-plane $\mathbb{R}^2_+$ $(z_2 > 0)$. This may be done by means of the following transformation "rectifying" the free boundary Γ_d:

$$x_1 = z_1, \qquad x_2 = z_2 + h(z_1)\chi\left(\frac{z}{|z|}\right). \tag{10.1.6}$$

Here $\chi(\xi)$ is a smooth function on the semicircle $S_+ : |\xi| = 1, \xi_2 > 0$, which is equal to one for $\xi \in S_+, \xi_1 > \xi_2/2$, and to zero for $\xi_1 < \xi_2/4$. Transformation (10.1.6) is invertible if $|h(z_1)\frac{(\partial}{\partial z_2)}\xi(z/|z|)| < 1$ which is the case if d is sufficiently small, and it establishes one-to-one correspondence between $\Gamma_d = \{x_2 = h(x_1),\ 0 < x_1 < d\}$ and the interval $\hat{\Gamma}_d = \{z_2 = 0,\ 0 < z_1 < d\}$ leaving Σ_d invariant. The Jacobi matrix of the inverse transformation J with the elements $J_{mk}, m, k = 1, 2$, has the form

$$J = \begin{pmatrix} 1 & 0 \\ -\dfrac{(h\xi)'_{z_1}}{1+h\xi'_{z_2}} & \dfrac{1}{1+h\xi'_{z_2}} \end{pmatrix}.$$

In the neighborhood of $\hat{\Gamma}_d$,

$$\chi = 1 \text{ and } J = \begin{pmatrix} 1 & 0 \\ -h'_{z_1} & 1 \end{pmatrix}.$$

In the new coordinates $\{z\}$, equations and boundary conditions (10.1.4), (10.1.5) take the form

$$\begin{aligned} &-\nu\hat{\nabla}^2\mathbf{w} - (\mathbf{a}\cdot\hat{\nabla})\mathbf{w} + (\mathbf{w}\cdot\hat{\nabla})\mathbf{w} + \hat{\nabla} p = 0, \qquad \hat{\nabla}\cdot\mathbf{w} = 0, \\ &\mathbf{w}|_{\Sigma_d} = 0, \qquad \mathbf{w}\cdot\mathbf{n}|_{\hat{\Gamma}_d} = an_1, \qquad \sigma H\mathbf{n} - \hat{T}(\mathbf{w}, p)\mathbf{n}|_{\hat{\Gamma}_d} = 0, \end{aligned} \tag{10.1.7}$$

$$\frac{\partial \mathbf{w}\cdot\mathbf{n}}{\partial\tau} - \frac{a}{\sigma}n_2\mathbf{n}\cdot\hat{T}(\mathbf{w}, p)\mathbf{n}|_{\hat{\Gamma}_d} = 0, \tag{10.1.8}$$

where

$$\hat{\nabla} = J^*\nabla = \left(\sum_{m=1}^{2} J_{mi}\frac{\partial}{\partial z_m}\right)_{i=1,2}, \qquad \hat{T}_{ik} = -p\delta_{ik} + \nu\hat{S}_{ik},$$

$$\hat{S}_{ik} = \sum_{m=1}^{2}\left(J_{mi}\frac{\partial w_j}{\partial z_m} + J_{mj}\frac{\partial w_i}{\partial z_m}\right),$$

$\mathbf{n}$ and τ are defined by (10.1.3) with $h = h(z_1)$, hence,

$$\frac{\partial}{\partial \tau} = \frac{1}{\sqrt{1+h'^2}}\frac{\partial}{\partial z_1}.$$

The image of Ω_d under transformation (10.1.6) contains a certain semi disk $|z| \le d_1,\ z_2 > 0$. Let $\zeta(z)$ be a smooth function equal to 1 for $|z| < d_1/2$ and to 0 for $|z| > d_1$. Clearly, we can consider $\mathbf{u} = \mathbf{w}\zeta$, $q = p\zeta$ as the solution of the following problem in the half-plane $\mathbb{R}^2_+$:

$$\begin{aligned} &-\nu\nabla^2\mathbf{u} + \nabla q = \mathbf{f}, \qquad \nabla\cdot\mathbf{u} = r,\\ &\mathbf{u}|_{\mathbb{R}_-} = 0, \qquad S_{12}(\mathbf{u})|_{\mathbb{R}_+} = b,\\ &\frac{\partial u_2}{\partial z_1} - \frac{a}{\sigma}(-p + \nu S_{22}(\mathbf{u}))|_{\mathbb{R}_+} = d. \end{aligned} \tag{10.1.9}$$

Here

$$\begin{aligned} &\mathbb{R}_- = \{z_2 = 0,\ z_1 < 0\}, \qquad \mathbb{R}_+ = \{z_2 = 0,\ z_1 > 0\},\\ &\mathbf{f} = \mathbf{f}_1 + \mathbf{f}_2, \quad r = r_1 + r_2, \quad b = b_1 + b_2, \quad d = d_1 + d_2, \qquad (10.1.10)\\ &\mathbf{f}_1 = -L(\mathbf{u}, q) + \{-\nu\nabla^2\mathbf{u} + \nabla q\}, \qquad \mathbf{f}_2 = L(\mathbf{w}\zeta, p\zeta) - \zeta L(\mathbf{w}, p),\\ &r_1 = (\nabla - \hat{\nabla})\mathbf{u}, \qquad r_2 = \hat{\nabla}\zeta\cdot\mathbf{w},\\ &b_1 = S_{12}(\mathbf{u}) - \sum_{i,j=1}^{2}\tau_i\hat{S}_{ij}(\mathbf{u})n_j,\\ &b_2 = \sum_{i,j,m=1}^{2}\tau_i\left(J_{mi}\frac{\partial\zeta}{\partial z_m}w_j + J_{mj}\frac{\partial\zeta}{\partial z_m}w_i\right)n_j,\\ &d_1 = \frac{\partial u_2}{\partial z_1} - \frac{1}{\sqrt{1+h'^2}}\frac{\partial\mathbf{u}\cdot\mathbf{n}}{\partial z_1} - \frac{a\nu}{\sigma}\left[\frac{p}{\nu}(n_2 - 1) + S_{22}(\mathbf{u}) - n_2\mathbf{n}\cdot\hat{S}\mathbf{n}\right],\\ &d_2 = \frac{1}{\sqrt{1+h'^2}}(\mathbf{w}\cdot\mathbf{n})\frac{\partial\zeta}{\partial z_1} - 2\frac{a\nu}{\sigma}(\mathbf{w}\cdot\mathbf{n})(\mathbf{n}\cdot\hat{\nabla}\zeta), \end{aligned}$$

and $L(\mathbf{u}, q) = -\nu\hat{\nabla}^2\mathbf{u} - (\mathbf{a}\cdot\hat{\nabla})\mathbf{u} + (\mathbf{u}\cdot\hat{\nabla})\mathbf{w} + \hat{\nabla}q$. All the functions in (10.1.9) are supposed to vanish for $|z| > d_1,\ z_2 > 0$.

We assume that the solution $(\mathbf{w}, p)$ of (10.1.9), (10.1.10) whose properties we are going to study belongs to a certain Hölder weighted space. Consider functions defined in $\Pi_g = \{|z| < g,\ z_2 > 0, g > 0\}$, and denote by $\overset{\circ}{C}{}^l_s(\Pi_g)$ $(l > 0,\ s \in \mathbb{R})$ the set of all functions $u(z),\ z \in \Pi_g$, with a finite norm

$$|u|_{\overset{\circ}{C}{}^l_s(\Pi_g)} = \sum_{0\le|j|<l} \sup_{x\in\Pi_g} |x|^{|j|-s}|D^j u(x)| + \sum_{|j|=[l]} \sup_{x\in\Pi_g} |x|^{l-s} \sup_{y\in\Pi_g:|x-y|\le|x|/2} |x-y|^{[l]-l}|D^j u(x) - D^j u(y)|. \quad (10.1.11)$$

In the same manner we can define spaces $\overset{\circ}{C}{}^l_s(\Sigma_g)$ and $\overset{\circ}{C}{}^l_s(\hat\Gamma_g)$. For instance, the norm in $\overset{\circ}{C}{}^l_s(\hat\Gamma_g)(\hat\Gamma_g = \{z_2 = 0,\ 0 < z_1 < g\})$ is given by

$$|u|_{\overset{\circ}{C}{}^l_s(\Gamma_g)} = \sum_{j=0}^{[l]} \sup_{x_1\in\hat\Gamma_g} x_1^{j-s}\left|\frac{d^j u}{dx_1^j}\right| + \sup_{x_1\in\hat\Gamma_g} x_1^{l-s} \sup_{y\in\hat\Gamma_g:|x-y|\le x_1/2} |x_1-y_1|^{[l]-l}\left|\frac{d^{[l]}u(x_1)}{dx_1^{[l]}} - \frac{d^{[l]}u(y_1)}{dy_1^{[l]}}\right|. \quad (10.1.12)$$

The main result of paper [1] is as follows:

Theorem 10.1.1. *Suppose that* $\mathbf{w} \in \overset{\circ}{C}{}^{2+\alpha}_\beta(\Gamma_{d_1}),\ p \in \overset{\circ}{C}{}^{1+\alpha}_{\beta-1}(\Gamma_{d_1}),\ \alpha \in (0,1),\ \beta \in (0, \frac{1}{2})$, *satisfy the equations and boundary conditions* (10.1.9), *and let* $h \in \overset{\circ}{C}{}^{3+\alpha}_{1+\beta}(\hat\Gamma_{d_1})$. *Then*

$$\mathbf{u} = a_0\mathbf{v}_0 + a_1\mathbf{v}_1 + \mathbf{u}',$$
$$q = a_0 p_0 + a_1 p_1 + q', \quad (10.1.13)$$

where $a_0,\ a_1$ *are some constants,* $(\mathbf{v}_0, p_0)$ *and* $(\mathbf{v}_1, p_1)$ *are special solutions of homogeneous problem* (10.1.9) *having the form*

$$\mathbf{v}_0(x) = \rho^{1/2}\mathbf{V}_0(\varphi), \qquad p_0 = \rho^{-1/2}q_0(\varphi),$$
$$\mathbf{v}_1(x) = \rho^{1-\kappa}\mathbf{V}_1(\varphi), \qquad p_1 = \rho^{-\kappa}q_1(\varphi),$$

with $\mathbf{V}_i,\ q_i(\varphi)$ *defined by formulas* (10.2.6), (10.2.7) *below. Here* (ρ,φ) *are standard polar coordinates on* $\mathbb{R}^2_+$, $\kappa \in (0, \frac{1}{2})$ *is a root of the equation*

$$\tan\kappa\pi = \frac{2a\nu}{\sigma}, \quad (10.1.14)$$

and finally $\mathbf{v}' \in \overset{\circ}{C}{}^{2+\alpha}_\gamma(\Gamma_{d_1}),\ p' \in \overset{\circ}{C}{}^{1+\alpha}_{\gamma-1}$ *with* $\gamma \in (1-\kappa, 1)$. *Moreover,*

$$h(x_1) = c_0 a_1 x_1^{2-\kappa} + \psi(x_1), \quad (10.1.15)$$

where $c_0 = -4\nu[\sigma\kappa(2-\kappa)]^{-1}$ *and* $\psi \in \overset{\circ}{C}{}^{3+\alpha}_{1+\gamma}(\hat\Gamma_{d_1})$.

In fact, the asymptotics of type (10.1.13) was established in [1] for the stream function $\Phi(u_1 = -\partial\Phi/\partial x_2,\ u_2 = \partial\Phi/\partial x_1)$, and the coordinates (z_1, z_2) in [1] were Mises' coordinates but Theorem 10.1.1 is equivalent to the main result of [1]. Formulas (10.1.13) follow from the well-known results of V.A. Kondrat'ev, V.G.Maz'ya, and B.A.Plamenevskii on the behavior of solutions of elliptic boundary value problems in neighborhoods of angular points at the boundary applied to problem (10.1.9).

We give here estimates for the coefficients $a_0,\ a_1$ assuming that the last boundary condition in (10.1.7) is inhomogeneous and has a form

$$\sigma H\mathbf{n} - \hat{T}(\mathbf{w}, p)\mathbf{n} = F\mathbf{n}. \tag{10.1.16}$$

The function F can appear as a result of the normalization of the pressure in which case it is a constant. It enters also into (10.1.8):

$$\frac{\partial \mathbf{w}\cdot\mathbf{n}}{\partial \tau} - \frac{a}{\sigma}n_2\mathbf{n}\cdot\hat{T}(\mathbf{w}, p)\mathbf{n} = \frac{a}{\sigma}n_2 F,$$

and into the function $d(x_1)$ from (10.1.9). It can be easily verified that the statement of Theorem 10.1.1 holds if $F \in \overset{\circ}{C}{}^{1+\alpha}_{\gamma-1}(\Gamma_{d_1})$. The main result of this paper is formulated below (Theorem 10.3.1). Our estimate for a_1 makes it possible to assert that the free boundary is located on one side of the line $x_2 = 0$ (in Fig. 10.1, above this line), at least for small values of the parameter a. This can be used in the proof of the solvability of some free boundary problems with moving contact points when a contact angle equals π. In this chapter only the problem of filling an infinite plane capillary and a piston problem are studied. Following a traditional scheme [2], [3] we find, at first, a solution with a free boundary Γ possibly leaving the capillary (i.e., the strip $|x| < R_0$) and establish a posteriori that Γ is located in this strip.

The condition that a contact angle equals π permits us to avoid the well-known difficulty consisting in the fact that the velocity vector field is discontinuous at a moving contact point and, what is even worse, it should have an infinite Dirichlet integral in any neighborhood of this point for arbitrary value of the contact angle in the interval $(0, \pi)$. Another way to avoid this difficulty is to replace the adherence boundary condition with the slip condition, at least near a moving contact point (this is done, for instance, in [4]–[6]). The problem of filling a plane capillary with a contact angle π is also studied by L.K. Antanovskii [7] with the help of the complex-valued representation of the solution of the Stokes equations. According to the scheme presented in [7], without rigorous justification, we can get the information on the coefficient a_1 studying the asymptotics of the free boundary as $a \to 0$.

This chapter was written when the author was working at the University of Paderborn as a visiting professor. The author is thankful to DFG (German Research Society) for the financial support.

10.2. On the Half-Plane Problem

Consider a homogeneous problem (10.1.9), i.e.,

$$
\begin{aligned}
&-\nu\nabla^2\mathbf{v}+\nabla p=0, \qquad \nabla\cdot\mathbf{v}=0, \qquad x\in\mathbb{R}^2_+,\\
&\mathbf{v}|_{\mathbb{R}_-}=0, \qquad \left.\frac{\partial v_1}{\partial x_2}+\frac{\partial v_2}{\partial x_1}\right|_{\mathbb{R}_+}=0,\\
&\frac{\partial v_2}{\partial x_1}+k\left(p-2\nu\frac{\partial v_2}{\partial x_2}\right)|_{\mathbb{R}_+}=0,
\end{aligned}
\tag{10.2.1}
$$

with $k=a/\sigma$, and find solutions to this problem of the form

$$
\mathbf{v}=\rho^\lambda\mathbf{V}(\varphi), \qquad p=\rho^{\lambda-1}q(\varphi),
$$

where (ρ,φ) are standard polar coordinates, so that $\mathbb{R}_+=\{\varphi=0\}$ and $\mathbb{R}_-=\{\varphi=\pi\}$. Let v and w be a radial and an angular component of $\mathbf{V}$, respectively, i.e.,

$$
v=V_1\cos\varphi+V_2\sin\varphi, \qquad w=-V_1\sin\varphi+V_2\cos\varphi.
$$

It is easy to verify (see [8], for example) that v,w,q should be a solution of the boundary value problem

$$
\begin{aligned}
&-v_{\varphi\varphi}+(1-\lambda^2)v+2w_\varphi+(\lambda-1)q\nu^{-1}=0,\\
&-w_{\varphi\varphi}+(1-\lambda^2)w-2v_\varphi+q_\varphi\nu^{-1}=0,\\
&w_\varphi+(\lambda+1)v=0 \qquad (0<\varphi<\pi),
\end{aligned}
\tag{10.2.2}
$$

$$
\begin{aligned}
&v(\pi)=w(\pi)=0, \qquad v_\varphi+(\lambda-1)w|_{\varphi=0}=0,\\
&\lambda w+k\nu(\nu^{-1}q-2w_\varphi-2v)|_{\varphi=0}=0.
\end{aligned}
\tag{10.2.3}
$$

System (10.2.2) has four linearly independent solutions (v,w,q). If $\lambda\neq 0$, they coincide with the columns of the matrix

$$
\begin{pmatrix}
(1-\lambda)\cos(1-\lambda)\varphi & \cos(1+\lambda)\varphi & (1-\lambda)\sin(1-\lambda)\varphi & \sin(1+\lambda)\varphi\\
-(1+\lambda)\sin(1-\lambda)\varphi & -\sin(1+\lambda)\varphi & (1+\lambda)\cos(1-\lambda)\varphi & \cos(1+\lambda)\varphi\\
-4\nu\lambda\cos(1-\lambda)\varphi & 0 & -4\nu\lambda\sin(1-\lambda)\varphi & 0
\end{pmatrix}.
$$

The solution of (10.2.2), (10.2.3) is a linear combination of these columns with the coefficients c_1-c_4. Boundary conditions (10.2.3) lead to a linear algebraic system for c_i, namely,

$$
\begin{pmatrix}
-(1-\lambda)\cos\lambda\pi & -\cos\lambda\pi & (1-\lambda)\sin\lambda\pi & -\sin\lambda\pi\\
-(1+\lambda)\sin\lambda\pi & \sin\lambda\pi & -(1+\lambda)\cos\lambda\pi & -\cos\lambda\pi\\
0 & 0 & -2\lambda(1-\lambda) & 2\lambda\\
-2\lambda k\nu(1+\lambda) & 2\lambda k\nu & \lambda(1+\lambda) & \lambda
\end{pmatrix}
\begin{pmatrix} c_1\\ c_2\\ c_3\\ c_4\end{pmatrix}=0.
$$

The determinant of this system equals $-8\lambda^2\cos^2\lambda\pi(\tan\lambda\pi+2k\nu)$. Additional investigation shows that problem (10.2.2), (10.2.3) with $\lambda=0$ has no nontrivial solutions. Indeed, a general solution of system (10.2.2) with $\lambda=0$ has a form

$$v = c_5 \cos\varphi + c_6 \sin\varphi + c_7\varphi\cos\varphi + c_8\varphi\sin\varphi,$$
$$w = (-c_5 - c_8)\sin\varphi + (c_6 - c_7)\cos\varphi + c_8\varphi\cos\varphi - c_7\varphi\sin\varphi,$$
$$q = -2c_8\nu\cos\varphi + 2c_7\nu\sin\varphi,$$

and it follows from (10.2.3) that $c_5 = c_6 = c_7 = c_8 = 0$.

Hence, the set of λ for which problem (10.2.2), (10.2.3) has nontrivial solutions consists of the roots of equations

$$\cos\lambda\pi = 0, \tag{10.2.4}$$

$$\tan\lambda\pi + 2k\nu = 0. \tag{10.2.5}$$

The corresponding (v, w, q) are

$$\begin{aligned} v(\varphi) &= -\frac{1-\lambda}{1+\lambda}\cos(1-\lambda)\varphi - \cos(1+\lambda)\varphi, \\ w(\varphi) &= \sin(1-\lambda)\varphi + \sin(1+\lambda)\varphi, \\ q(\varphi) &= \frac{4\lambda\nu}{1+\lambda}\cos(1-\lambda)\varphi, \end{aligned} \tag{10.2.6}$$

and

$$\begin{aligned} v(\varphi) &= \cos(1-\lambda)\varphi - \cos(1+\lambda)\varphi - \tan\lambda\pi[\sin(1-\lambda)\varphi + \sin(1+\lambda)\varphi], \\ w(\varphi) &= -\frac{1+\lambda}{1-\lambda}\sin(1-\lambda)\varphi + \sin(1+\lambda)\varphi \\ &\quad - \tan\lambda\pi\left[\frac{1+\lambda}{1-\lambda}\cos(1-\lambda)\varphi + \cos(1+\lambda)\varphi\right], \\ q(\varphi) &= -\frac{4\lambda\nu}{1-\lambda}\cos(1-\lambda)\varphi + \frac{4\lambda\nu}{1-\lambda}\tan\lambda\pi\sin(1-\lambda)\varphi, \end{aligned} \tag{10.2.7}$$

respectively.

Functions (10.2.6) also satisfy the condition $w(0) = 0$ and, as a consequence,

$$\frac{q}{\nu} - 2w_\varphi - 2v|_{\varphi=0} = 0. \tag{10.2.8}$$

Positive roots of (10.2.4), (10.2.5) can be numbered in an increasing order. The minimal positive root is $\lambda_0 = \frac{1}{2}$ and the following two are $\lambda_1 = 1 - \kappa \in (\frac{1}{2}, 1)$ and $\lambda_2 = \frac{3}{2}$. The functions (v_0, w_0, q_0) and (v_1, w_1, q_1) are given by (10.2.6) and (10.2.7) with $\lambda = \lambda_0$ and $\lambda = \lambda_1$, respectively.

We shall also consider solutions of a "dual" homogeneous problem (10.2.1) with $-k$ instead of k in boundary conditions. The corresponding λ are roots of equations

$$\cos\lambda\pi = 0, \tag{10.2.9}$$

$$\tan \lambda\pi - 2k\nu = 0, \tag{10.2.10}$$

and v, w, q are given by the same formulas (10.2.6), (10.2.7).

Let us now consider a nonhomogeneous problem (10.1.9) with arbitrary $\mathbf{f}$, r, b, d. This is an elliptic boundary value problem in an infinite angle with the opening π and with the vertex placed at the origin. On the sides of the angle, $\mathbb{R}_-$ and $\mathbb{R}_+$, there are prescribed different boundary conditions. We can apply to this problem the following theorem due to V.A. Kondratiev [9], V.G. Maz'ya and B.A. Plamenevskii [10],[11].

Theorem 10.2.1.

(a) *Let l be an arbitrary positive noninteger and let $s < l+2$ be different from the roots of* (10.2.4) *or* (10.2.5). *For arbitrary $\mathbf{f} \in \overset{\circ}{C}{}^{l}_{s-2}(\mathbb{R}^2_+)$, $r \in \overset{\circ}{C}{}^{l+1}_{s-1}(\mathbb{R}^2_+)$, $b \in \overset{\circ}{C}{}^{l+1}_{s-1}(\mathbb{R}_+)$, $d \in \overset{\circ}{C}{}^{l+1}_{s-1}(\mathbb{R}_+)$ problem* (10.1.9) *has a unique solution $\mathbf{u} \in \overset{\circ}{C}{}^{l+2}_{s}(\mathbb{R}^2_+)$, $q \in \overset{\circ}{C}{}^{l+1}_{s-1}(\mathbb{R}^2_+)$, and*

$$|\mathbf{u}|_{\overset{\circ}{C}{}^{l+2}_{s}(\mathbb{R}^2_+)} + |q|_{\overset{\circ}{C}{}^{l+1}_{s-1}(\mathbb{R}^2_+)} \le$$
$$c_9(|\mathbf{f}|_{\overset{\circ}{C}{}^{l}_{s-2}(\mathbb{R}^2_+)} + |r|_{\overset{\circ}{C}{}^{l+1}_{s-1}(\mathbb{R}^2_+)} + |b|_{\overset{\circ}{C}{}^{l+1}_{s-1}(\mathbb{R}_+)} + |d|_{\overset{\circ}{C}{}^{l+1}_{s-1}(\mathbb{R}_+)})$$

with a constant c_9 depending on s and l.

(b) *Let (s,l) and (s',l') satisfy the hypotheses* (a) *and let $s' > s$. If $\mathbf{f} \in \overset{\circ}{C}{}^{l}_{s-2}(\mathbb{R}^2_+) \cap \overset{\circ}{C}{}^{l'}_{s'-2}(\mathbb{R}^2_+)$, $r \in \overset{\circ}{C}{}^{l+1}_{s-1}(\mathbb{R}^2_+) \cap \overset{\circ}{C}{}^{l'+1}_{s'-1}(\mathbb{R}^2_+)$, $b \in \overset{\circ}{C}{}^{l+1}_{s-1}(\mathbb{R}_+) \cap \overset{\circ}{C}{}^{l'+1}_{s'-1}(\mathbb{R}_+)$, $d \in \overset{\circ}{C}{}^{l+1}_{s-1}(\mathbb{R}_+) \cap \overset{\circ}{C}{}^{l'+1}_{s'-1}(\mathbb{R}_+)$, then the corresponding solutions $(\mathbf{u}, q) \in \overset{\circ}{C}{}^{l+2}_{s}(\mathbb{R}^2_+) \times \overset{\circ}{C}{}^{l+1}_{s-1}(\mathbb{R}^2_+)$ and $(\mathbf{u}', q') \in \overset{\circ}{C}{}^{l'+2}_{s'}(\mathbb{R}^2_+) \times \overset{\circ}{C}{}^{l'+1}_{s'-1}(\mathbb{R}^2_+)$ are related to each other as follows:*

$$\mathbf{u} = \sum_n a_n \mathbf{u_n}(x) + \mathbf{u}'(x),$$
$$q = \sum_n a_n q_n(x) + q'(x). \tag{10.2.11}$$

Here $\mathbf{v}_n = \rho^{\lambda_n}\mathbf{V}_n(\varphi)$, $p_n = \rho^{\lambda_n - 1} q_n(\varphi)$ are special solutions of the homogeneous problem (10.2.1) *constructed above, λ_n are roots of* (10.2.4) *or* (10.2.5), *$\mathbf{V}_n(\varphi) = (v_n, w_n)$, and q_n are corresponding solutions of problem* (10.2.2), (10.2.3) *given by* (10.2.6) *or* (10.2.7), *respectively, with $\lambda = \lambda_n$. The summation is extended to all λ_n in the interval*

$$s < \lambda_n < s'.$$

Now we are going to derive explicit formulas for the coefficients a_n in (10.2.11). Let $(\mathbf{u}_n, s_n)$ be solutions of a "dual" homogeneous problem (10.2.1) with $-k$ instead of k in boundary conditions. They have a form

$$\mathbf{u}_n = \rho^{\hat{\lambda}_n}\mathbf{U}_n(\varphi), \qquad s_n = \rho^{\hat{\lambda}_n - 1} t_n(\varphi),$$

where $\hat{\lambda}_n$ are roots of equations (10.2.9), (10.2.10). We consider only solutions with negative $\hat{\lambda}_n$ which are numbered in decreasing order. Cleary, $\hat{\lambda}_n = -\lambda_n$, so that $\hat{\lambda}_0 = -\frac{1}{2}$, $\hat{\lambda}_1 = -1 + \kappa$, etc.

Theorem 10.2.2 *The coefficients a_n in* (10.2.11) *are determined by*

$$a_n = \frac{1}{A_n}\left[\int_0^\infty \left(\nu b u_{n1} - \frac{1}{k} d u_{n2}\right) d\rho + \int_{\mathbb{R}^2_+} r(x) s_n(x),\ dx \right.$$
$$\left. - \int_{\mathbb{R}^2_+} (\mathbf{f} - \nu\nabla r)\cdot \mathbf{u}_n\, dx \right], \tag{10.2.12}$$

where

$$A_n = \frac{1}{k} w_n(0)\hat{w}_n(0) - \int_o^\pi [T(\mathbf{v}_n, p_n)\mathbf{n}\cdot\mathbf{u}_n - T(\mathbf{u}_n, s_n)\mathbf{n}\cdot\mathbf{v}_n]|_{\rho=1}\, d\varphi,$$
$$A_n = \frac{8\pi\nu\lambda_n}{1-\lambda_n^2}, \tag{10.2.13}$$

if λ_n is a root of (10.2.4), *and*

$$A_n = \frac{8\pi\nu\lambda_n}{(1-\lambda_n^2)\cos^2\lambda_n\pi}, \tag{10.2.14}$$

if λ_n is a root of (10.2.5); *$\hat{w}_n$ is an angular component of $\mathbf{U}_n$.*

Proof. We write the Green's identity

$$\int_{\mathbb{R}_\epsilon} (-\nu\nabla S(\mathbf{u}) + \nabla q)\cdot\mathbf{u}_m\, dx - \int_{\mathbb{R}_\epsilon} (-\nu\nabla^2\mathbf{u}_m + \nabla s_m)\cdot\mathbf{u}\, dx$$
$$= \int_{\mathbb{R}_\epsilon} s_m \nabla\cdot\mathbf{u}\, dx - \int_{S_\epsilon} [-T(\mathbf{u}, q)\mathbf{n}\cdot\mathbf{u}_m + T(\mathbf{u}_m, s_m)\mathbf{n}\cdot\mathbf{u}]\, dS$$
$$+ \int_\epsilon^\infty [T(\mathbf{n}, q)\mathbf{n}\cdot\mathbf{n}_m - T(\mathbf{n}_m, s_m)\mathbf{n}\cdot\mathbf{u}]|_{x_2=0}\, dx_1, \tag{10.2.15}$$

where $\mathbb{R}_\epsilon = \{x_2 > 0,\ |x| > \epsilon\}$, $S_\epsilon = \{x_2 > 0, |x| = \epsilon\}$, $\mathbf{n} = \mathbf{x}|x|^{-1}$ on S_ϵ, and $\mathbf{n} = (0, 1)$ on $\mathbb{R}_+$. All the integrals in (10.2.15) are convergent, since the integrands decay at infinity sufficiently rapidly. From (10.1.9) it follows that

$$\int_{\mathbb{R}_\epsilon} (\mathbf{f} - \nu\nabla r)\cdot\mathbf{u}_m\, dx - \int_{\mathbb{R}_\epsilon} r s_m\, dx - \int_\epsilon^\infty \left(\nu b u_{m1} - \frac{1}{k} d u_{m2}\right) d\rho$$
$$= \frac{1}{k}\int_\epsilon^\infty \left(\frac{\partial u_2}{\partial x_1} u_{m2} + \frac{\partial u_{m2}}{\partial x_1} u_2\right)_{x_2=0} dx_1$$
$$+ \int_{S_\epsilon} [T(\mathbf{u}, q)\mathbf{n}\cdot\mathbf{u}_m - T(\mathbf{u}_m, s_m)\mathbf{n}\cdot\mathbf{u}]\, dS. \tag{10.2.16}$$

Let us pass to the limit as $\epsilon \to 0$. We substitute (10.2.11) into the right-hand side of (10.2.16) and observe that all the terms containing $\mathbf{u}'$ and q' vanish in the limit because $\lambda_m < s'$. Moreover, if $n \neq m$, then

$$-\frac{1}{k}v_{n2}(\epsilon,0)u_{m2}(\epsilon,0) + \int_{S_\epsilon} [T(\mathbf{v}_n, p_n)\mathbf{n}\cdot\mathbf{u}_m - T(\mathbf{u}_m, s_m)\mathbf{n}\cdot\mathbf{v}_n]\,dS$$
$$= -\epsilon^{\lambda_n-\lambda_m}A_{nm} = 0,$$

where

$$A_{nm} = \frac{1}{k}w_n(0)\hat{w}_m(0) - \int_0^\pi [T(\mathbf{v}_n, p_n)\mathbf{n}\cdot\mathbf{u}_m - T(\mathbf{u}_m, s_m)\mathbf{n}\cdot\mathbf{v}_n]|_{\rho=1}\,d\varphi$$
$$= 0. \tag{10.2.17}$$

Hence, passing to the limit in (10.2.16) we arrive at (10.2.12).

It remains to verify (10.2.17) and to compute A_n. Let $M >> 1 >> \epsilon$ and let $\mathbb{R}_{\epsilon,M} = \mathbb{R}_\epsilon \setminus \mathbb{R}_M$. We have

$$0 = \int_{\mathbb{R}_{\epsilon,M}} (-\nu\nabla^2\mathbf{v}_n + \nabla p_n)\cdot\mathbf{u}_m\,dx - \int_{\mathbb{R}_{\epsilon,M}} (-\nu\nabla^2\mathbf{u}_m + \nabla s_m)\cdot\mathbf{v}_n\,dx$$
$$= -\int_{S_M} [T(\mathbf{v}_n, p_n)\mathbf{n}\cdot\mathbf{u}_m - T(\mathbf{u}_m, s_m)\mathbf{n}\cdot\mathbf{v}_n]\,dS$$
$$+\int_{S_\epsilon} [T(\mathbf{v}_n, p_n)\mathbf{n}\cdot\mathbf{u}_m - T(\mathbf{u}_m, s_m)\mathbf{n}\cdot\mathbf{v}_n]\,dS$$
$$+\frac{1}{k}\int_\epsilon^M \left(\frac{\partial v_{n2}}{\partial x_1}u_{m2} + \frac{\partial u_{m2}}{\partial x_1}v_{n2}\right)_{x_2=0} dx_1 = (M^{\lambda_n-\lambda_m} - \epsilon^{\lambda_n-\lambda_m})A_{nm},$$

which proves (10.2.17). In order to compute A_n, we write the Green's identity for $(\mathbf{v}_n, p_n)$ and $(\mathbf{u} = \rho^\lambda\mathbf{U}(\varphi),\ s = \rho^{\lambda-1}q(\varphi))$ where λ is close to $-\lambda_n$ and $q(\varphi)$ and the components of $\mathbf{U}(\varphi)$ are given by formulas (10.2.6) or (10.2.7). Suppose that λ_n is a root of (10.2.4). Since

$$\mathbf{u}|_{\varphi=\pi} = \left(\frac{2}{1+\lambda}\cos\lambda\pi, 0\right)\rho^\lambda,$$
$$u_2|_{\varphi=0} = s - 2\nu\frac{\partial v_2}{\partial x_2}\bigg|_{\varphi=0} = 0, \qquad \frac{\partial u_2}{\partial x_1} + \frac{\partial u_1}{\partial x_2} = 0,$$

this identity takes the form

$$0 = \int_{\mathbb{R}_{\epsilon,M}} (-\nu\nabla^2\mathbf{v}_n + \nabla p_n)\cdot\mathbf{u}\,dx - \int_{\mathbb{R}_{\epsilon,M}} (-\nu\nabla^2\mathbf{u} + \nabla s)\cdot\mathbf{v}_n\,dx$$
$$= \frac{1}{k}\int_\epsilon^M \left(\frac{\partial v_{n2}}{\partial x_1}u_2 + \frac{\partial u_2}{\partial x_1}v_{n2}\right)_{x_2=0} dx_1 - \int_{S_M} [T(\mathbf{v}_n, p_n)\mathbf{n}\cdot\mathbf{u} - T(\mathbf{u}, s)\cdot\mathbf{v}_n]\,dS$$
$$+\int_{S_\epsilon} [T(\mathbf{v}_n, p_n)\mathbf{n}\cdot\mathbf{u} - T(\mathbf{u}, s)\mathbf{n}\cdot\mathbf{v}_n]\,dS + \int_{-M}^{-\epsilon} T(\mathbf{v}_n, p_n)\mathbf{n}\cdot\mathbf{u}|_{x_2=0}\,dx_1,$$

or

$$(M^{\lambda_n+\lambda} - \epsilon^{\lambda_n+\lambda}) \left(\frac{1}{k} w_n(0)\hat{w}(0) - \int_0^{\pi} [T(\mathbf{v}_n, p_n)\mathbf{n}\cdot\mathbf{u} - T(\mathbf{u}, s)\mathbf{n}\cdot\mathbf{v}_n]\, d\varphi \right)$$

$$= -\int_{-M}^{-\epsilon} T(\mathbf{v}_n, p_n)\mathbf{n}\cdot\mathbf{u}|_{x_2=0}\, dx_1 = -\int_{\epsilon}^{M} T(\mathbf{v}_n, p_n)\mathbf{n}\cdot\mathbf{u}|_{\varphi=\pi}\, d\rho$$

$$= -\frac{2\nu\cos\lambda\pi}{1+\lambda}(v_{n\varphi} + (\lambda_n - 1)w_n)|_{\varphi=\pi}\frac{1}{\lambda+\lambda_n}(M^{\lambda+\lambda_n} - \epsilon^{\lambda+\lambda_n}), \quad (10.2.18)$$

where $\hat{w}$ is the angular component of $\mathbf{U}$. If we cancel $M^{\lambda+\lambda_n} - \epsilon^{\lambda+\lambda_n}$ and let λ tend to $-\lambda_n$ we arrive at

$$A_n = -\frac{2\nu}{1-\lambda_n}(v_{n\varphi} + (\lambda_n - 1)w_n)|_{\varphi=\pi} \lim_{\lambda\to-\lambda_n} \frac{\cos\lambda\pi}{\lambda+\lambda_n}$$

$$= \frac{8\lambda_n\pi\nu}{1-\lambda_n^2}\sin^2\lambda_n\pi = \frac{8\lambda_n\pi\nu}{1-\lambda_n^2}.$$

In a similar way we can compute A_n when λ_n is a root of (10.2.5). In this case, $\mathbf{u}$ satisfies the following boundary conditions:

$$\mathbf{u}|_{\varphi=\pi} = 0, \qquad \left.\frac{\partial u_1}{\partial x_2} + \frac{\partial u_2}{\partial x_1}\right|_{\varphi=0} = 0,$$

$$\left.\frac{\partial u_2}{\partial x_1} - k\left(s - 2\nu\frac{\partial u_2}{\partial x_2}\right)\right|_{\varphi=0} = -\rho^{\lambda-1}\frac{2\lambda}{1-\lambda}(\tan\lambda\pi - 2k\nu),$$

and instead of (10.2.18) we have

$$0 = (M^{\lambda_n+\lambda} - \epsilon^{\lambda_n+\lambda})\left\{\frac{1}{k}w_n(0)\hat{w}(0) - \int_0^{\pi}[T(\mathbf{v}_n, p_n)\mathbf{n}\cdot\mathbf{u} - T(\mathbf{u}, s)\mathbf{n}\cdot\mathbf{v}_n]|_{\rho=1}\, dS\right\}$$

$$+ \frac{1}{k}\int_{\epsilon}^{\pi} v_{n2}\rho^{\lambda-1}|_{\varphi=0}\, d\rho\frac{2\lambda}{1-\lambda}(\tan\lambda\pi - 2k\nu).$$

The last term is equal to

$$-\frac{M^{\lambda_n+\lambda} - \epsilon^{\lambda_n+\lambda}}{k(\lambda_n+\lambda)}\tan\lambda_n\pi\frac{2}{1-\lambda_n}\frac{2\lambda}{1-\lambda}(\tan\lambda\pi - 2k\nu),$$

so at the end we get

$$A_n = -\frac{4\lambda_n}{1-\lambda_n^2}\frac{\tan\lambda_n\pi}{k}\lim_{\lambda\to-\lambda_n}\frac{\tan\lambda\pi - 2k\nu}{\lambda_n+\lambda} = \frac{8\pi\nu\lambda_n}{(1-\lambda_n^2)\cos^2\lambda_n\pi}.$$

The theorem is proved.

Theorems 10.2.1 and 10.2.2 hold both for positive and for negative a. In this chapter we assume that $a > 0$, and we make use of Theorem 10.1.1 for $l = l' = \alpha \in (0, 1)$, $s = \beta \in (0, \frac{1}{2})$, $s' = \gamma \in (1-\kappa, 1)$. Of course, this theorem remains true in the case of nonhomogeneous boundary conditions on $\mathbb{R}_- : \mathbf{v} = \mathbf{c} \in \overset{\circ}{C}{}^{2+\alpha}_{\beta}(\mathbb{R}_-)$.

We observe at the conclusion that explicit formulas for coefficients in the asymptotics of solutions of elliptic boundary value problems in neighborhoods of angular and conical points were obtained by V.G. Maz'ya and B.A. Plamenevskii (see [11], for example).

10.3. Asymptotics of the Solution of Problem (10.1.9) Near the Origin

Let us turn to problem (10.1.9) with functions $\mathbf{f}, r, b, d$ specified in (10.1.10) and below. We observe first of all that Theorem 10.1.1 can be proved with the help of Theorem 10.2.1 according to the following scheme. Since $h \in \overset{\circ}{C}{}^{3+\alpha}_{1+\beta}(\hat{\Gamma}_{d_1})$, the transformation (10.1.6) differs from the identical transformation by a mapping of class $\overset{\circ}{C}{}^{3+\alpha}_{1+\beta}(\Pi_{d_1})$, and $J_{kj} - \delta_{kj} \in \overset{\circ}{C}{}^{2+\alpha}_{\beta}(\hat{\Gamma}_{d_1})$. It is straightforward to show that $\mathbf{f} \in \overset{\circ}{C}{}^{\alpha}_{2\beta-2}(\Pi_{d_1})$, $r \in \overset{\circ}{C}{}^{1+\alpha}_{2\beta-1}(\Pi_{d_1})$, $b, d \in \overset{\circ}{C}{}^{1+\alpha}_{2\beta-1}(\hat{\Gamma}_{d_1})$, and

$$|\mathbf{f}|_{\overset{\circ}{C}{}^{\alpha}_{2\beta-2}(\Pi_{d_1})} + |r|_{\overset{\circ}{C}{}^{1+\alpha}_{2\beta-1}(\Pi_{d_1})} + |b|_{\overset{\circ}{C}{}^{1+\alpha}_{2\beta-1}(\hat{\Gamma}_{d_1})} + |d|_{\overset{\circ}{C}{}^{1+\alpha}_{2\beta}(\hat{\Gamma}_{d_1})}$$
$$\leq c_1 \left(|\mathbf{w}|_{\overset{\circ}{C}{}^{2+\alpha}_{\beta}(\Pi_{d_1})} + |p|_{\overset{\circ}{C}{}^{1+\alpha}_{\beta-1}(\Pi_{d_1})} \right).$$

Hence, if $2\beta < \frac{1}{2}$, then, by Theorem 10.2.1, $\mathbf{u} \in \overset{\circ}{C}{}^{2+\alpha}_{2\beta}(\Pi_{d_1})$ and $q \in \overset{\circ}{C}{}^{1+\alpha}_{2\beta-1}(\Pi_{d_1})$. Now we can show that $\mathbf{f} \in \overset{\circ}{C}{}^{\alpha}_{3\beta-2}(\Pi_{d_1})$, $r \in \overset{\circ}{C}{}^{1+\alpha}_{3\beta-1}(\Pi_{d_1})$, $b, d \in \overset{\circ}{C}{}^{1+\alpha}_{3\beta-1}(\hat{\Gamma}_{d_1})$, etc. Repeating these arguments we arrive sooner or later at the conclusion that $\mathbf{u} \in \overset{\circ}{C}{}^{2+\alpha}_{\beta_1}(\Pi_{d_1})$, $q \in \overset{\circ}{C}{}^{1+\alpha}_{\beta_1-1}(\Pi_{d_1})$, and $\mathbf{f} \in \overset{\circ}{C}{}^{\alpha}_{\beta_2-2}(\Pi_{d_1})$, $r \in \overset{\circ}{C}{}^{1+\alpha}_{\beta_2-1}(\Pi_{d_1})$, $b, d \in \overset{\circ}{C}{}^{1+\alpha}_{\beta_2-1}(\hat{\Gamma}_{d_1})$ with $\beta_1 < \frac{1}{2} < \beta_2 < 1-\kappa$, and the sum of norms of all these functions is estimated by

$$c_2(|\mathbf{w}|_{\overset{\circ}{C}{}^{2+\alpha}_{\beta}(\Pi_{d_1})} + |p|_{\overset{\circ}{C}{}^{1+\alpha}_{\beta-1}(\Pi_{d_1})}).$$

Further, by virtue of Theorem 10.2.1,

$$\mathbf{u} = a_0 \mathbf{v}_0 + \mathbf{u}'', \qquad q = a_0 p_0 + q'', \tag{10.3.1}$$

where $\mathbf{u}'' \in \overset{\circ}{C}{}^{2+\alpha}_{\beta_2}(\mathbb{R}^2_+)$, $q'' \in \overset{\circ}{C}{}^{1+\alpha}_{\beta_2-1}(\mathbb{R}^2_+)$ is also a solution of problem (10.1.9), hence,

$$|\mathbf{u}''|_{\overset{\circ}{C}{}^{2+\alpha}_{\beta_2}(\mathbb{R}^2_+)} + |q''|_{\overset{\circ}{C}{}^{1+\alpha}_{\beta_2-1}(\mathbb{R}^2_+)} \leq c_3(|\mathbf{f}|_{\overset{\circ}{C}{}^{\alpha}_{\beta_2-2}(\Pi_{d_1})} + |r|_{\overset{\circ}{C}{}^{1+\alpha}_{\beta_2}(\Pi_{d_1})}$$
$$+ |b|_{\overset{\circ}{C}{}^{1+\alpha}_{\beta_2-2}(\hat{\Gamma}_{d_1})} + |d|_{\overset{\circ}{C}{}^{1+\alpha}_{\beta_2-1}(\hat{\Gamma}_{d_1})})$$
$$\leq c_2 c_3(|\mathbf{w}|_{\overset{\circ}{C}{}^{2+\alpha}_{\beta}(\Pi_{d_1})} + |p|_{\overset{\circ}{C}{}^{1+\alpha}_{\beta-1}(\Pi_{d_1})}).$$

Consider now the equation for h:

$$-\sigma\frac{h''}{(1+h'^2)^{3/2}} = \mathbf{n}\cdot\hat{T}(\mathbf{w},p)\mathbf{n}|_{x_2=h(x_1)}$$
$$= a_0\mathbf{n}\cdot\hat{T}(\mathbf{v}_0,p_0)\mathbf{n}+\mathbf{n}\cdot\hat{T}(\mathbf{u}'',q'')\mathbf{n}|_{x_2=h(x_1)}, \quad (10.3.2)$$

where $\mathbf{n}$ is defined by (10.1.3). By virtue of (10.2.8), $\mathbf{v}_0$ and p_0 satisfy the condition

$$T_{22}(\mathbf{v}_0,p_0)|_{z_2=0} = -p_0+2\nu\frac{\partial v_{02}}{\partial z_2}|_{z_2=0}=0,$$

hence, on the line $z_2=0$

$$\mathbf{n}\cdot\hat{T}(\mathbf{v}_0,p_0)\mathbf{n} = \mathbf{n}\cdot[\hat{T}(\mathbf{v}_0,p_0)-T(\mathbf{v}_0,p_0)]\mathbf{n}+\nu[\mathbf{n}\cdot S(\mathbf{v}_0)\mathbf{n}-S_{22}(\mathbf{v}_0)]$$
$$\in \overset{\circ}{C}{}^{1+\alpha}_{\beta_3-1}(\hat{\Gamma}_{d1}),$$

with $\beta_3=\frac{1}{2}+\beta>\frac{1}{2}$. As $\mathbf{n}\cdot\hat{T}(\mathbf{u}'',q'')\mathbf{n}\in\overset{\circ}{C}{}^{1+\alpha}_{\beta_2-1}(\hat{\Gamma}_{d_1})$, we conclude from (10.3.2) that $h\in\overset{\circ}{C}{}^{3+\alpha}_{\beta_4+1}(\hat{\Gamma}_{d_1})$, $\beta_4=\min(\beta_2,\beta_3)>\frac{1}{2}$, hence, $J_{kj}-\delta_{kj}\in\overset{\circ}{C}{}^{2+\alpha}_{\beta_4}(\Pi_{d_1})$. Turning again to problem (10.1.9) we show that $\mathbf{f}\in\overset{\circ}{C}{}^{\alpha}_{\beta_4-3/2}(\Pi_{d_1})$, $r\in\overset{\circ}{C}{}^{1+\alpha}_{\beta_4-1/2}(\Pi_{d_1})$, $b,d\in\overset{\circ}{C}{}^{1+\alpha}_{\beta_4-1/2}(\hat{\Gamma}_{d_1})$, and

$$|\mathbf{f}|_{\overset{\circ}{C}{}^{\alpha}_{\beta_4-3/2}(\Pi_{d_1})}+|r|_{\overset{\circ}{C}{}^{1+\alpha}_{\beta_4-1/2}(\Pi_{d_1})}+|b|_{\overset{\circ}{C}{}^{1+\alpha}_{\beta_4-1/2}(\hat{\Gamma}_{d_1})}+|d|_{\overset{\circ}{C}{}^{1+\alpha}_{\beta_4-1/2}(\hat{\Gamma}_{d_1})}$$
$$\le c_4(|\mathbf{w}|_{\overset{\circ}{C}{}^{2+\alpha}_{\beta}(\Pi_{d_1})}+|p|_{\overset{\circ}{C}{}^{1+\alpha}_{\beta-1}(\Pi_{d_1})}). \quad (10.3.3)$$

Since $\beta_4>\frac{1}{2}$, we obtain (10.1.13) by applying Theorem 10.2.1 once more. To prove (10.1.15), we consider again equation (10.3.2) and write it in the form

$$-\sigma\frac{h''}{(1+h'^2)^{3/2}} = a_1T_{22}(\mathbf{v}_1,p_1)+[a_1(\mathbf{n}\cdot\hat{T}(\mathbf{v}_1,p_1)\mathbf{n}-T_{22}(\mathbf{v}_1,p_1))$$
$$+a_0\mathbf{n}\cdot\hat{T}(\mathbf{v}_0,p_0)\mathbf{n}]+\mathbf{n}\cdot\hat{T}(\mathbf{u}',q')\mathbf{n}|_{\hat{\Gamma}_{d_1}}.$$

We know already that $J_{kj}-\delta_{kj}\in\overset{\circ}{C}{}^{2+\alpha}_{\beta_4}(\Pi_{d_1})$ which makes it possible to show that the expression in square brackets belongs to $\overset{\circ}{C}{}^{1+\alpha}_{\beta_4-1/2}(\hat{\Gamma}_{d_1})$. This yields (10.1.15) with

$$c_0=\frac{\nu}{\sigma\lambda_1(1+\lambda_1)}\left(\frac{q}{\nu}-2w_{1\varphi}-2v_1\right)\bigg|_{\varphi=0} = -\frac{4\nu}{\sigma(1-\lambda_1^2)} = -\frac{4}{\sigma\kappa(2-\kappa)}.$$

We are now ready to prove one of the main results of this chapter.

Theorem 10.3.1. *Suppose that* $\mathbf{w}$ *and* p *satisfy the relations* (10.1.7) *but the last boundary condition has a form* (10.1.16) *with* $F\in\overset{\circ}{C}{}^{1+\alpha}_{\gamma-1}(\hat{\Gamma}_{d_1})$, $\gamma\in(1-\kappa,1)$. *Then*

$$a_1 = \frac{a\kappa \cos^2 \kappa\pi}{2\pi\sigma(1+\kappa)} \int_0^\infty \zeta(x_1, 0) F(x_1) x_1^{\kappa-1}\, dx_1 + a_1', \tag{10.3.4}$$

where

$$|a_1'| \leq c_5 a (a + |\mathbf{w}|_{\overset{o}{C}{}^{2+\alpha}_{\beta}(\Pi_{d_1})} + |p|_{\overset{o}{C}{}^{1+\alpha}_{\beta-1}(\Pi_{d_1})}) \tag{10.3.5}$$

and the constant c_5 is independent of a.

Proof. The coefficient a_1 is given by the formula (10.2.12); we only have to add $kn_2F\zeta$ to the function d. This gives the contribution

$$\begin{aligned}
-\frac{1}{A_1}\int_0^\infty \zeta F n_2 u_{12}\, dx_1 &= -\frac{2\tan\lambda_1\pi}{(1+\lambda_1)A_1}\int_0^\infty \zeta F n_2 x_1^{\kappa-1}\, dx_1 \\
&= \frac{a\kappa\cos^2\kappa\pi}{2\pi\sigma(1-\kappa)}\left(\int_0^\infty \zeta F x_1^{\kappa-1} dx_1 + \int_0^\infty F\zeta(n_2-1)x_1^{\kappa-1}\, dx_1\right).
\end{aligned}$$

Clearly,

$$\frac{a\kappa\cos^2\kappa\pi}{2\pi\sigma(1-\kappa)}\int_0^\infty |F\zeta(n_2-1)| x_1^{\kappa-1}\, dx_1 \leq c_6 a^2$$

with a constant c_6 independent of a. Other terms in (10.2.12) are evaluated in the following way:

the first term vanishes since $u_{n1}|_{\varphi=0} = 0$,

$$\begin{aligned}
\left|\frac{1}{k}\int_0^\infty du_{n2}\, dx_1\right| &\leq c_7 \int_0^{d_1} |d(x_1)| x_1^{\kappa-1}\, dx_1 \\
&\leq c_8 |d|_{\overset{o}{C}{}^{1+\alpha}_{\beta_4-1/2}(\hat{\Gamma}_{d_1})} \int_0^{d_1} x_1^{\beta_4-1/2-1+\kappa}\, dx_1 \leq c_9 |d|_{\overset{o}{C}{}^{1+\alpha}_{\beta_4-1/2}(\hat{\Gamma}_{d_1})},
\end{aligned}$$

$$\begin{aligned}
&\left|\int_{\mathbb{R}^2_+} r s_n\, dx\right| + \left|\int_{\mathbb{R}^2_+} (\mathbf{f} - \nu\nabla r)\cdot \mathbf{u}_n\, dx\right| \\
&\quad \leq c_{10}(|r|_{\overset{o}{C}{}^{1+\alpha}_{\beta_4-1/2}(\Pi_{d_1})} + |\mathbf{f}|_{\overset{o}{C}{}^{\alpha}_{\beta_4-3/2}(\Pi_{d_1})}) \int_{\Pi_{d_1}} |x|^{\beta_4-1/2-\kappa-2} dx \\
&\quad \leq c_{11}(|r|_{\overset{o}{C}{}^{1+\alpha}_{\beta_4-1/2}(\Pi_{d_1})} + |\mathbf{f}|_{\overset{o}{C}{}^{\alpha}_{\beta_4-3/2}(\Pi_{d_1})}).
\end{aligned}$$

Constants $c_7 - c_{11}$ are independent of κ since $\beta_4 > \frac{1}{2}$. Taking (10.3.3) into account (with the additional term $c_{12}a$ on the right-hand side) we arrive at (10.3.5). The theorem is proved. ■

Corollary. *If $F \leq -F_0 < 0$ and $a + |\mathbf{w}|_{\overset{o}{C}{}^{2+\alpha}_{\beta}(\Pi_{d_1})} + |q|_{\overset{o}{C}{}^{1+\alpha}_{\beta-1}(\Pi_{d_1})}$ is small in comparison with F_0, then $a_1 < 0$ and $c_0 a_1 > 0$.*

Indeed,

$$a_1 < -\frac{aF_0\kappa\cos^2\kappa\pi}{2\pi\sigma(1-\kappa)}\int_0^{d_1/2} x_1^{\kappa-1}dx_1 + |a_1'| < -\frac{aF_0\cos^2\kappa\pi}{2\pi\sigma(1-\kappa)}\left(\frac{d_1}{2}\right)^{\kappa} + |a_1'| < -c_{13}a,$$

and

$$c_0 a_1 > \frac{4\nu}{\sigma\kappa(1-\kappa)}\left(\frac{aF_0\cos^2\kappa\pi}{2\pi\sigma(1-\kappa)} - |a_1'|\right) > c_{14},$$

where c_{13} and c_{14} are independent of a.

Theorem 10.3.2. *Suppose that* $F - F(0) \in \overset{\circ}{C}{}^{1+\alpha}_{s-1}(\hat{\Gamma}_{d_1})$, $s > 1$. *Then the remainders in* (10.1.13), (10.1.15) *have the properties*

$$\mathbf{u}' \in \overset{\circ}{C}{}^{2+\alpha}_{s_1}(\Pi_{d_1}), \qquad q' = F(0) + r', \qquad r' \in \overset{\circ}{C}{}^{1+\alpha}_{s_1-1}(\Pi_{d_1}), \qquad \psi \in \overset{\circ}{C}{}^{3+\alpha}_{s_1+1}(\hat{\Gamma}_{d_1}),$$

with $s_1 = \min(s, \beta_4 + \frac{1}{2}) > 1$, *and*

$$|\mathbf{u}'|_{\overset{\circ}{C}{}^{2+\alpha}_{s_1}(\Pi_{d_1})} + |r'|_{\overset{\circ}{C}{}^{1+\alpha}_{s_1-1}(\Pi_{d_1})} \leq c_{15}\left(|\mathbf{w}|_{\overset{\circ}{C}{}^{2+\alpha}_{\beta}(\Pi_{d_1})} + |p|_{\overset{\circ}{C}{}^{1+\alpha}_{\beta-1}(\Pi_{d_1})} + a\right). \quad (10.3.6)$$

Proof. Let $\zeta_\epsilon(x) = \zeta_1(\epsilon x)$ where $\zeta_1 \in C_0^\infty(\mathbb{R}^2)$, $\zeta_1(x) = 1$ for $|x| < 1$, $\zeta_1(x) = 0$ for $|x| > 2$. It is easy to see that $\mathbf{u}' \in \overset{\circ}{C}{}^{2+\alpha}_{\gamma}(\mathbb{R}^2_+)$, $r' = q' - F(0)\zeta_\epsilon \in \overset{\circ}{C}{}^{1+\alpha}_{\gamma-1}(\mathbb{R}^2_+)$ is a solution to the problem

$$\begin{aligned} -\nu\nabla^2\mathbf{u}' + \nabla r' &= \mathbf{f} + \mathbf{f}', \qquad \nabla\cdot\mathbf{u}' = r', \\ \mathbf{u}'|_{\mathbb{R}_-} &= 0, \quad S_{12}(\mathbf{u}')|_{\mathbb{R}_+} = b, \\ \frac{\partial u_2'}{\partial z_1} + k(r' - \nu S_{22}(\mathbf{u}'))|_{\mathbb{R}_+} &= d + d', \end{aligned}$$

with $\mathbf{f}' = -F(0)\nabla\zeta_\epsilon \in \overset{\circ}{C}{}^{\alpha}_{\gamma-2}(\mathbb{R}^2_+) \cap \overset{\circ}{C}{}^{\alpha}_{s_1-2}(\mathbb{R}^2_+)$, $d' = \kappa(\zeta n_2 F - \zeta\epsilon F(0)) \in \overset{\circ}{C}{}^{1+\alpha}_{\gamma-1}(\mathbb{R}_+) \cap \overset{\circ}{C}{}^{1+\alpha}_{s_1-1}(\mathbb{R}_+)$. Hence, by Theorem 10.2.1, $\mathbf{u}' \in \overset{\circ}{C}{}^{2+\alpha}_{s_1}(\mathbb{R}^2_+)$, $r' \in \overset{\circ}{C}{}^{1+\alpha}_{s_1-1}(\mathbb{R}^2_+)$.

Consider the equation for h:

$$\begin{aligned} -\frac{\sigma h''}{(1+h'^2)^{3/2}} &= a_1 T_{22}(\mathbf{v}_1, p_1) + [a_1(\mathbf{n}\cdot\hat{T}(\mathbf{v}_1, p_1)\mathbf{n}) \\ &\quad - T_{22}(\mathbf{v}_1, p_1)) + a_0\mathbf{n}\cdot\hat{T}(\mathbf{v}_0, p_))\mathbf{n}] \\ &\quad + \mathbf{n}\cdot\hat{T}(\mathbf{u}', q')\mathbf{n} + F, \qquad x \in \hat{\Gamma}_{d_1}. \end{aligned}$$

Since

$$\mathbf{n}\cdot\hat{T}(\mathbf{u}', q')\mathbf{n} + F = \mathbf{n}\cdot\hat{T}(\mathbf{u}', r')\mathbf{n} + (F - F(0)) \in \overset{\circ}{C}{}^{1+\alpha}_{s_1-1}(\hat{\Gamma}_{d_1}),$$

this equation implies (10.1.15) with $\psi \in \overset{\circ}{C}{}^{3+\alpha}_{s_1+1}(\hat{\Gamma}_{d_1})$. Estimate (10.3.6) follows from Theorem 10.2.1 and from (10.3.3) (with additional term $c_{12}a$ on the right-hand side), if we choose $\epsilon \leq a^{\frac{1}{s_1-1}}$. The theorem is proved. ∎

Formula (10.1.15) with $c_0 a > c_{14}$ and $\psi \in \overset{\circ}{C}{}^{3+\alpha}_{s_1+1}(\hat{\Gamma}_{d_1})$ shows that the curve Γ is located above x_1-axis in a certain neighborhood of the origin which is independent of a.

Formulas (10.1.13), (10.1.15) also hold for $a < 0$, provided that $\beta \in (|\kappa|, \frac{1}{2})$. In this case $\kappa \in (-\frac{1}{2}, 0)$, and it may be shown that $c_0 a_1 > 0$. The case of negative a will be considered in one of the forthcoming papers.

10.4. Free Boundary Problem with Moving Contact Points

Consider the problem of quasi-stationary filling a plane capillary, i.e., the strip $|x_1| < R_0$, $-\infty < x_2 < \infty$. The liquid occupies an infinite domain Ω bounded by the lines $x_1 = \pm R_0$ and by a free surface Γ which is moving with a constant speed $a > 0$ in the direction of the x_2-axis. Ω is located under Γ, i.e., from the side of smaller values of x_2. A total flux across an arbitrary cross-section $x_2 = \text{const}$ of Ω is prescribed and it is equal to $2R_0 a$, and a contact angle equals π. In the coordinate system moving together with the free surface, this problem becomes stationary, and it can be written in the form

$$\begin{gathered} -\nu\nabla^2\mathbf{w} - (\mathbf{a}\cdot\nabla)\mathbf{w} + (\mathbf{w}\cdot\nabla)\mathbf{w} + \nabla p = 0, \qquad \nabla\cdot\mathbf{w} = 0, \qquad x \in \Omega, \\ \mathbf{w}|_\Sigma = 0, \qquad \mathbf{w}\cdot\mathbf{n} - an_2|_\Gamma = 0, \qquad \sigma H\mathbf{n} - T(\mathbf{w}, p)\mathbf{n}|_\Gamma = 0, \\ \int_{-R_0}^{R_0} w_2(x)\,dx_1 = 2R_0 a, \end{gathered} \tag{10.4.1}$$

where $\mathbf{a} = (0, a)$ and $\mathbf{n}$ is a unit exterior normal to Γ. The free boundary Γ meets the walls $x_1 = \pm R_0$ at the points $x_\pm = (\pm R_0, 0)$, so that $\Sigma = \{x_2 < 0,\ x_1 = \pm R_0\}$, and $n_2(x_\pm) = 0$.

Let us normalize the pressure by the requirement

$$\int_\Gamma (\mathbf{n}\cdot T(\mathbf{w}, p)\mathbf{n})n_2\,dS = 0 \tag{10.4.2}$$

and separate the tangential and the normal component in the dynamical boundary condition on Γ. Then this condition takes the form

$$\tau\cdot S(\mathbf{w})\mathbf{n}|_\Gamma = 0, \tag{10.4.3}$$

$$\sigma H - \mathbf{n}\cdot T(\mathbf{w}, p)\mathbf{n}|_\Gamma = F, \tag{10.4.4}$$

where τ is a tangential vector to Γ with the components $\tau_1 = n_2$, $\tau_2 = -n_1$. The constant F may be computed with the help of (10.4.2) in the following way:

$$F\int_\Gamma n_2\,dS = \sigma\int_\Gamma Hn_2\,dS = \sigma\int_\Gamma \frac{\partial\tau_2}{\partial\tau}dS = -2\sigma.$$

Since $\int_\Gamma n_2\,dS = 2R_0$, it follows that

$$-F = \frac{\sigma}{R_0} \equiv -F_0. \tag{10.4.5}$$

We will look for the solution of problem (10.4.1) in weighted Hölder spaces. For the infinite domain Ω we define spaces $\overset{\circ}{C}{}^l_s(\Omega)$ as sets of functions given in Ω and having a finite norm (10.1.11) with Π_g replaced by Ω and with the weight function $|x|$ changed for $W(x) = \min(|x-x_+|, |x-x_-|, 1)$ (this means that the singular set consists of two points $x_\pm$ and that the weight does not play any role for large $|x|$). The curve Γ will be defined by the equation

$$|x| = R(\varphi), \qquad \varphi\in(0,\pi)\equiv I, \tag{10.4.6}$$

where φ is a standard polar angle on the plane $\mathbb{R}^2$ (so that the polar coordinates of $x_\pm$ are $(R_0, 0)$ and (R_0,π), respectively). We will consider functions given on Γ as functions of $\varphi\in I$, and we define $\overset{\circ}{C}{}^l_s(I)$ as the space of functions $f(\varphi)$, $\varphi\in I$, with a finite norm

$$|f|_{\overset{\circ}{C}{}^l_s(I)} = \sum_{j=0}^{[l]}\sup_{\varphi\in I} w^{j-s}(\varphi)|\frac{d^j f(\varphi)}{d\varphi^j}| + [f]_{\overset{\circ}{C}{}^l_s(I)},$$

$$[f]_{\overset{\circ}{C}{}^l_s(I)} = \sup_{\varphi\in I} w^{l-s}(\varphi)\sup_{\substack{\varphi'\in I\\ |\varphi-\varphi'|\le\frac12 w(\varphi)}} |\varphi-\varphi'|^{-l+[l]}\left|\frac{d^{[l]}f(\varphi)}{d\varphi^{[l]}} - \frac{d^{[l]}f(\varphi')}{d\varphi'^{[l]}}\right|,$$

where $w(\varphi) = \min(\varphi, \pi-\varphi)$.

For $s\in(0,l)$, we also introduce spaces $C^l_s(I)$ whose elements need not vanish at the endpoints $\varphi = 0$ and $\varphi=\pi$ of I, as spaces with norms

$$|f|_{C^l_s(I)} = \sum_{s<j<l}\sup_{\varphi\in I} w^{j-s}(\varphi)\left|\frac{d^j f(\varphi)}{d\varphi^j}\right| + [f]_{C^l_s(I)} + |f|_{C^s(I)},$$

where

$$|f|_{C^s(I)} = \sum_{j=0}^{[s]}\sup_{\varphi\in I}\left|\frac{d^j f}{d\varphi^j}\right| + \sup_{\varphi,\varphi'\in I}|\varphi-\varphi'|^{-s+[s]}\left|\frac{d^{[s]}f(\varphi)}{d\varphi^{[s]}} - \frac{d^{[s]}f(\varphi')}{d\varphi'^{[s]}}\right|$$

is a usual Hölder norm. For $s<0$ we set $C^l_s(I) = \overset{\circ}{C}{}^l_s(I)$. In an analogous way we define $C^l_s(\Omega)$ (see general definitions in [8]). Finally, by $\overset{\circ}{C}{}^l_{s,+}(I)$, $C^l_{s,+}(I)$ and $\overset{\circ}{C}{}^l_{s,-}(I)$, $C^l_{s,-}(I)$ we mean the subspaces of $\overset{\circ}{C}{}^l_s(I)$ or $C^l_s(I)$ whose elements are even or odd functions of $\pi/2-\varphi$, respectively. If Ω is symmetric with respect to the x_2-axis, we may define $\overset{\circ}{C}{}^l_{s,+}(\Omega)$ and $\overset{\circ}{C}{}^l_{s,-}(\Omega)$ as the subspaces of $\overset{\circ}{C}{}^l_s(\Omega)$ consisting of even or odd functions of x_1, and also the spaces $C^l_{s,+}(\Omega)$ and $C^l_{s,-}(\Omega)$. We say that $\Gamma \in C^l_s$ if the function $R(\varphi)$ in (10.4.6) belongs to $C^l_s(I)$. In terms of $R(\varphi)$ the curvature H of Γ is defined by

$$H[R] \equiv \frac{1}{R}\frac{d}{d\varphi}\frac{R_\varphi}{\sqrt{R^2+R_\varphi^2}} - \frac{1}{\sqrt{R^2+R_\varphi^2}},$$

hence, (10.4.4) may be written in the form

$$\sigma H[R] = \mathbf{n}\cdot T(\mathbf{w},p)\mathbf{n}|_{|x|=R(\varphi)} + F. \tag{10.4.7}$$

The radial and angular component of the exterior normal $\mathbf{n}$ is equal to $R(\varphi)/\sqrt{R^2+R_\varphi^2}$ and $-R_\varphi/\sqrt{R^2+R_\varphi^2}$, respectively.

The main result of this section is as follows:

Theorem 10.4.1. *For small $a>0$, problem* (10.4.1), (10.4.2) *has a solution* $\mathbf{w}, p, \Gamma$ *with the following properties:*

$$\mathbf{w} \in \overset{o}{C}{}^{2+\alpha}_{\beta}(\Omega),\qquad \nabla p \in \overset{o}{C}{}^{\alpha}_{\beta-2}(\Omega),\qquad \Gamma \in C^{3+\alpha}_{1+\beta},\qquad \alpha\in(0,1),\qquad \beta\in(0,\tfrac{1}{2}).$$

The curve Γ which is determined by equation (10.4.6) *has endpoints* $(\pm R_0, 0)$ *(hence, $R(0) = R(\pi) = R_0$) and is located in the strip $-R_0 < x < R_0$. The solution possesses the symmetry properties*

$$\mathbf{w}(-x_1,x_2) = (-w_1(x), w_2(x)),\qquad p(-x_1,x_2) = p(x),\qquad R(\pi-\varphi) = R(\varphi).$$

As $|x|\to\infty$, $\mathbf{w}$ and p tend to the functions corresponding to the Poiseuille flow, i.e.,

$$\mathbf{w} \to \left(0, \tfrac{3}{2}a\left(1-\frac{x_1^2}{R_0^2}\right)\right),\qquad \nabla p \to \left(0, -\frac{3a\nu}{R_0^2}\right),$$

and finally there holds the estimate

$$|\mathbf{w}|_{\overset{o}{C}{}^{2+\alpha}_{\beta}(\Omega)} + |\nabla p|_{\overset{o}{C}{}^{\alpha}_{\beta-2}(\Omega)} + |R-R_0|_{\overset{o}{C}{}^{3+\alpha}_{1+\beta}(I)} \le c_1 a. \tag{10.4.8}$$

In the class of functions satisfying all these conditions, the solution is unique.

The idea of the proof is this: first of all, we construct the solution without paying attention to the condition $|x_1| < R_0$ for $x\in\Gamma$. Then from our previous considerations we conclude that $R - R_0 \in \overset{o}{C}{}^{3+\alpha}_{\gamma+1}(I)$ with $\gamma > \frac{1}{2}$ and that formulas (10.1.15), (10.3.4) are valid in the neighborhoods of $x_\pm$. For small a, the estimate (10.4.8) and the relation (10.1.15) guarantee that Γ lies in the strip $|x_1| < R_0$. The construction of the solution can be carried out by the procedure going back to the papers of V.V. Pukhnachov (see [2]). We need to investigate two linear problems. Given a symmetric $\Gamma \in C^{3+\alpha}_{1+\beta}$, $\alpha\in(0,1)$, $\beta\in(0,\frac{1}{2})$ with endpoints

$x_\pm = (\pm R_0, 0)$ not necessarily contained in the strip $|x_1| < R_0$; we consider the boundary value problem

$$\begin{aligned} &-\nu\nabla^2\mathbf{w} + \nabla s = \mathbf{f}, \qquad \nabla\cdot\mathbf{w} = r(x), \qquad x \in \Omega, \\ &\mathbf{w}|_\Sigma = 0, \qquad \mathbf{v}\cdot\mathbf{n}|_\Gamma = b, \qquad \tau\cdot S(\mathbf{w})\mathbf{n}|_\Gamma = d, \\ &\lim_{x_2\to-\infty}\int_{-R_0}^{R_0} v_2\,dx_1 = 0, \end{aligned} \tag{10.4.9}$$

with additional condition (10.4.2) in the domain Ω with $\partial\Omega = \Gamma\cup\Sigma$.

Theorem 10.4.2. *Suppose that* Γ *is defined by equation* (10.4.6) *with* $R - R_0 \in \overset{\circ}{C}{}^{3+\alpha}_{1+\beta}(I)$, $\alpha \in (0,1)$, $\beta \in (0, \frac{1}{2})$. *For arbitrary* $\mathbf{f} \in \overset{\circ}{C}{}^{\alpha}_{\beta-2}(\Omega)$, $r \in \overset{\circ}{C}{}^{1+\alpha}_{\beta-1}\Omega)$ *with a compact support and for arbitrary* $b \in \overset{\circ}{C}{}^{2+\alpha}_{\beta}(\Gamma)$, $d \in \overset{\circ}{C}{}^{1+\alpha}_{\beta-1}(\Gamma)$ *satisfying the necessary compatibility condition*

$$\int_\Gamma b\,dS = \int_\Omega r\,dx$$

problem (10.4.9),(10.4.2) *has a unique solution* $\mathbf{w} \in \overset{\circ}{C}{}^{2+\alpha}_{\beta}(\Omega)$, $s \in \overset{\circ}{C}{}^{1+\alpha}_{\beta-1}(\Omega)$. *As* $x_2 \to -\infty$, $\mathbf{w}(x) \to 0$, *and* $s(x) \to s_0 = const$ *exponentially. The solution satisfies the inequality*

$$\begin{aligned} &|\mathbf{w}|_{\overset{\circ}{C}{}^{2+\alpha}_{\beta}(\Omega)} + |s|_{\overset{\circ}{C}{}^{1+\alpha}_{\beta-1}(\Omega)} \\ &\quad\le c_2(|\mathbf{f}|_{\overset{\circ}{C}{}^{\alpha}_{\beta-2}(\Omega)} + |r|_{\overset{\circ}{C}{}^{1+\alpha}_{\beta-1}(\Omega)} + |b|_{\overset{\circ}{C}{}^{2+\alpha}_{\beta}(\Gamma)} + |d|_{\overset{\circ}{C}{}^{1+\alpha}_{\beta-1}(\Gamma)}). \end{aligned} \tag{10.4.10}$$

If Γ *is symmetric with respect to the* x_2*-axis, and if* $f_1 \in \overset{\circ}{C}{}^{\alpha}_{\beta-2,-}(\Omega)$, $f_2 \in \overset{\circ}{C}{}^{\alpha}_{\beta-2,+}(\Omega)$, $r \in \overset{\circ}{C}{}^{1+\alpha}_{\beta-1,+}(\Omega)$, $b \in \overset{\circ}{C}{}^{2+\alpha}_{\beta,+}(\Gamma)$, $d \in \overset{\circ}{C}{}^{1+\alpha}_{\beta-1,+}(\Gamma)$, *then* $w_1 \in \overset{\circ}{C}{}^{2+\alpha}_{\beta,-}(\Omega)$, $w_2 \in \overset{\circ}{C}{}^{2+\alpha}_{\beta,+}(\Omega)$, $s \in \overset{\circ}{C}{}^{1+\alpha}_{\beta-1,+}(\Omega)$.

The theorem can be proved by the methods developed in [12]–[14]. First of all, the problem is solved in the class of vector fields with the Dirichlet integrals decaying at infinity:

$$\int_{-t}^{-t+1} dx_2 \int_{-R_0}^{R_0} |\nabla\mathbf{w}|^2\,dx_1 \le ce^{-\varkappa t}, \qquad \varkappa > 0.$$

This estimate is established by the "technics of the Saint-Venant's principle" (see [12]). The estimates of the Hölder norms are then found from local considerations; they are consequences of regularity theorems for solutions of elliptic boundary value problems. The condition $\beta < \frac{1}{2}$ appears as a restriction of regularity near the angular points $x_\pm$ (see [8]).

The second linear problem we have to deal with is more elementary:

$$r_{\varphi\varphi} + r = g(\varphi), \qquad \varphi \in I, \qquad r_\varphi(0) = r_\varphi(\pi) = 0. \tag{10.4.11}$$

Theorem 10.4.3. *For arbitrary* $g \in \overset{\circ}{C}{}^{1+\alpha}_{\beta+1,+}(I)$, *problem* (10.4.11) *has a unique solution* $r \in C^{3+\alpha}_{1+\beta,+}(I)$ *given by the formula*

$$r(\varphi) = \int_0^{\varphi} \sin(\varphi - \xi) g(\xi)\, d\xi + \tfrac{1}{2} \cos\varphi \int_0^{\pi} \sin\xi g(\xi)\, d\xi. \quad (10.4.12)$$

If, in addition,

$$\int_0^{\pi} \sin\xi g(\xi)\, d\xi = 0, \quad (10.4.13)$$

then $r \in \overset{\circ}{C}{}^{3+\alpha}_{1+\beta,+}(I)$ *i.e., additional boundary conditions* $r(0) = r(\pi) = 0$ *are satisfied and*

$$|r|_{\overset{\circ}{C}{}^{3+\alpha}_{1+\beta}(I)} \le c_3 |g|_{\overset{\circ}{C}{}^{1+\alpha}_{\beta-1}(I)}. \quad (10.4.14)$$

Proof. It is straightforward to verify that the function (10.4.12) is even in $\pi/2 - \varphi$, belongs to $C^{3+\alpha}_{1+\beta}(I)$, and solves (10.4.11). Clearly, there is no other even solution. To prove (10.4.14), we observe that under condition (10.4.13)

$$|r(\varphi)||\varphi|^{-1} + |r_\varphi(\varphi)| \le c_4 \int_0^{\varphi} |g(\xi)|\, d\xi \le c_5 |\varphi|^{\beta} |g|_{\overset{\circ}{C}{}^{1+\alpha}_{\beta-1}(I)},$$

and from the symmetry properties of r it also follows that

$$|r(\varphi)||\pi - \varphi|^{-1} + |r_\varphi(\varphi)| \le c_6 |\pi - \varphi|^{\beta} |g|_{\overset{\circ}{C}{}^{1+\alpha}_{\beta-1}(I)}.$$

Higher-order derivatives of r are estimated from equation $r_{\varphi\varphi} = g - r$. The theorem is proved. ∎

We are now ready to prove Theorem 10.4.1.

Proof of Theorem 10.4.1. As we have already pointed out, it is enough to find the solution of our problem with $\Gamma = \{|x| = R(\varphi)\}$, $R - R_0 \in \overset{\circ}{C}{}^{3+\alpha}_{1+\beta,+}(I)$, not necessarily contained in the strip $|x_1| < R_0$. Let there be given a symmetric curve Γ defined by (10.4.6) with a small $R - R_0 \in \overset{\circ}{C}{}^{3+\alpha}_{1+\beta,+}(I)$. Consider an auxiliary problem

$$\begin{gathered} -\nu\nabla^2 \mathbf{w} - (\mathbf{a}\cdot\nabla)\mathbf{w} + (\mathbf{w}\cdot\nabla)\mathbf{w} + \nabla p = 0, \qquad \nabla\cdot\mathbf{w} = 0, \qquad x \in \Omega, \\ \mathbf{w}|_\Sigma = 0, \qquad \mathbf{w}\cdot\mathbf{n} - a n_2|_\Gamma = 0, \qquad \tau\cdot S(\mathbf{w})\mathbf{n}|_\Gamma = 0, \quad (10.4.15) \\ \int_{-R_0}^{R_0} w_2(x)\, dx_1 = 2R_0 a. \end{gathered}$$

In the case of small a it can be easily shown that this problem has a unique solution $(\mathbf{w}, p)$ in the form

$$\mathbf{w} = \mathbf{w}_0 \psi(x_2) + \mathbf{u}, \qquad p = p_0 \psi(x_2) + s,$$

where $(\mathbf{w}_0, p_0)$ correspond to the Poiseuille flow, i.e.,

$$\mathbf{w}_0 = \left(0, \frac{3a}{2}\left(1 - \frac{x_1^2}{R_0^2}\right)\right), \qquad p_0 = -\frac{3a\nu}{R_0^2}x_2,$$

$\psi(x_2)$ is a smooth monotone function of x_2 vanishing for $x_2 > -1$ and equal to 1 for $x_2 < -2$, and $(\mathbf{u}, s)$ is a solution to the problem

$$-\nu\nabla^2\mathbf{u} - (\mathbf{a}\cdot\nabla)\mathbf{u} + (\mathbf{w}_0\psi\cdot\nabla)\mathbf{u} + (\mathbf{u}\cdot\nabla)(\mathbf{w}_0\psi) + (\mathbf{u}\cdot\nabla)\mathbf{u} + \nabla s = \mathbf{f},$$
$$\nabla\cdot\mathbf{u} = r, \qquad x\in\Omega,$$
$$\mathbf{u}|_\Sigma = 0, \qquad \mathbf{u}\cdot\mathbf{n} - an_2|_\Gamma = 0, \qquad \tau\cdot S(\mathbf{u})\mathbf{n}|_\Gamma = 0,$$
$$\int_{-R_0}^{R_0} u_2(x)\,dx_1 = 0, \quad x_1 < -2, \qquad \int_\Gamma \mathbf{n}\cdot T(\mathbf{u}, s)\mathbf{n}n_2\,dS = 0.$$

Here $\mathbf{f}, r$ are defined by

$$r = -\mathbf{w}_0\cdot\nabla\psi = -w_{02}\psi'(x_2),$$
$$\mathbf{f} = -p_0\nabla\psi(x_2) + \nu\mathbf{w}_0\psi''(x_2) + a\psi'(x_2)\mathbf{w}_0 - w_{02}\psi(x_2)\psi'(x_2)\mathbf{w}_0,$$

so they are smooth functions vanishing for $x_2 < -2$, $x_2 > -1$.

If a is small, then the unique solvability of the auxiliary problem follows from Theorem 10.4.2, and $(\mathbf{w}, p)$ satisfy the inequality

$$|\mathbf{w}|_{\overset{\circ}{C}{}^{2+\alpha}_\beta(\Omega)} + |\nabla p|_{\overset{\circ}{C}{}^{\alpha}_{\beta-2}(\Omega)} \le c_8 a. \tag{10.4.16}$$

The whole problem now reduces to the equation (10.4.7) where $\mathbf{n}\cdot T\mathbf{n}$ should be regarded as a function depending on Γ, i.e., on R. Consider an equivalent equation

$$\sigma H + \frac{\sigma}{R_0} = \mathbf{n}\cdot T(\mathbf{w}, p)\mathbf{n} + F - F_0,$$

where F is a constant close to $F_0 = -\sigma R_0^{-1}$. If we multiply this equation by R and separate out terms linear with respect to $r = R - R_0$, we obtain

$$\frac{\sigma}{R_0}\left(\frac{d^2}{d\varphi^2}r + r\right) = Rg(\varphi) + G(r, r_\varphi, r_{\varphi\varphi}) \tag{10.4.17}$$

where

$$G(r, r_\varphi, r_{\varphi,\varphi}) = \sigma\left(\frac{d}{d\varphi}r_\varphi h(r, r_\varphi) - rh - R_0\left(h - \frac{r}{R_0^2}\right)\right),$$
$$h(r, r_\varphi) = \frac{1}{R_0} - \frac{1}{\sqrt{R^2 + r_\varphi^2}} = \frac{r_\varphi^2 + (R + R_0)r}{R_0\sqrt{R^2 + r_\varphi^2}(R_0 + \sqrt{R^2 + r_\varphi^2})},$$
$$g(\varphi) = \mathbf{n}\cdot T(\mathbf{w}, p)\mathbf{n} + F - F_0.$$

We are looking for a solution to (10.4.17) from the space $\overset{\circ}{C}{}^{3+\alpha}_{1+\beta,+}(I)$, i.e., satisfying the boundary conditions

$$r_\varphi(0) = r_\varphi(\pi) = 0, \qquad r(0) = r(\pi) = 0. \tag{10.4.18}$$

It is possible to find such a solution if and only if

$$\int_0^\pi \sin\varphi[Rg(\varphi) + G(r, r_\varphi, r_{\varphi\varphi})]\, d\varphi = 0,$$

i.e.,

$$F_0 - F = \left(2R_0 + \int_0^\pi r(\varphi)\sin\varphi\, d\varphi\right)^{-1} \int_0^\pi \sin\varphi[R\mathbf{n}\cdot T\mathbf{n}|_{|x|=R(\varphi)} + G]\, d\varphi. \tag{10.4.19}$$

Let us write (10.4.17)–(10.4.19) in the form

$$r = \frac{R_0}{\sigma}\mathcal{L}(Rg + G) \equiv \mathcal{M}r, \tag{10.4.20}$$

where $\mathcal{L}$ is the operator which assigns the solution $r \in \overset{\circ}{C}{}^{3+\alpha}_{1+\beta,+}(I)$ to problem (10.4.11) to every function $g \in \overset{\circ}{C}{}^{1+\alpha}_{\beta-1,+}(I)$ satisfying condition (10.4.13). The right-hand side of (10.4.20) determines a nonlinear operator $\mathcal{M}$ in the ball $\mathcal{B}_\delta$:

$$|r|_{\overset{\circ}{C}{}^{3+\alpha}_{1+\beta}(I)} \le \delta$$

of the space $\overset{\circ}{C}{}^{3+\alpha}_{1+\beta,+}(I)$ with a small radius δ. Our aim is to show that $\mathcal{M}$ is a contraction operator. It is obvious that for arbitrary $r, r' \in \mathcal{B}_\delta$ there hold inequalities

$$G(r, r_\varphi, r_{\varphi\varphi})|_{\overset{\circ}{C}{}^{1+\alpha}_{\beta-1}(I)} \le c_9\delta|r|_{\overset{\circ}{C}{}^{3+\alpha}_{1+\beta}(I)},$$
$$G(r, r_\varphi, r_{\varphi\varphi}) - G(r', r'_\varphi, r'_{\varphi\varphi})|_{\overset{\circ}{C}{}^{1+\alpha}_{\beta-1}(I)} \le c_{10}\delta|r - r'|_{\overset{\circ}{C}{}^{3+\alpha}_{1+\beta}(I)},$$

and if $r \in \overset{\circ}{C}{}^{3+\alpha}_{1+\beta,+}(I)$, then $G(r, r_\varphi, r_{\varphi\varphi}) \in \overset{\circ}{C}{}^{1+\alpha}_{\beta-1,+}(I)$. Consider now the function $t[r] = \mathbf{n}\cdot T\mathbf{n}|_\Gamma \in \overset{\circ}{C}{}^{1+\alpha}_{\beta-1,+}(I)$, $\Gamma = \{|x| = R_0 + r(\varphi)\}$, satisfying condition (10.4.2). It follows from the estimate (10.4.16) that

$$|t[r]|_{\overset{\circ}{C}{}^{1+\alpha}_{\beta-1}(I)} \le c_{11}a. \tag{10.4.21}$$

In order to compare $t[r]$ and $t[r']$, we need to construct a mapping $Y : \Omega \to \Omega'$ with the following properties:

(1) Y is an invertible mapping of the class $C^{3+\alpha}_{1+\beta}(\Omega)$ which is identical in the domain $\{x \in \Omega :\ x_2 < -1\}$, and $Yx = x|x|^{-1}(R_0 + r'(\varphi))$ for $x \in \Gamma$.

(2) The Jacobi matrix $J = \left(\frac{\partial z_i}{\partial x_k}\right)_{i,k=1,2} \equiv (J_{ik})$ satisfies the inequality

$$|J - I|_{\overset{o}{C}{}^{2+\alpha}_{\beta}(\Omega)} \le c_{12}|r - r'|_{\overset{o}{C}{}^{3+\alpha}_{1+\beta}(I)} \tag{10.4.22}$$

which means, in particular, that $J(x_{\pm}) = I$.

Such a mapping will be constructed at the end of the proof.

In the coordinates $z = Yx \in \Omega'$ problem (10.4.15) takes the form

$$\begin{aligned} &-\nu\nabla'^2\mathbf{w} - (\mathbf{a}\cdot\nabla')\mathbf{w} + (\mathbf{w}\cdot\nabla')\mathbf{w} + \nabla' p = 0, \qquad \nabla'\cdot\mathbf{w} = 0, \\ &\mathbf{w}|_{\Sigma} = 0, \qquad \mathbf{w}\cdot J^*\mathbf{n}' - a(J^*\mathbf{n}')_2|_{\Gamma'} = 0, \qquad \hat{J}\tau'\cdot S'(\mathbf{w})J^*\mathbf{n}'|_{\Gamma'} = 0. \end{aligned} \tag{10.4.23}$$

Here J^* and $\hat{J}$ are transposed and the associated matrix to J, respectively, i.e., $J^*_{ik} = J_{ki}$, $\hat{J}_{11} = J_{22}$, $\hat{J}_{22} = J_{11}$, $\hat{J}_{12} = -J_{12}$, $\hat{J}_{21} = -J_{21}$, and by τ' and $\mathbf{n}'$ we have denoted the tangential and the normal vector to $\Gamma' = \{|x| = R_0 + r'(\varphi)\}$. It can easily be verified that $\mathbf{n} = J^*\mathbf{n}'/|J^*\mathbf{n}'|$, $\tau = \hat{J}\tau'/|\hat{J}\tau'|$. Finally,

$$\nabla' = J^*\nabla = \left(\sum_{k=1}^{2} J_{ki}\frac{\partial}{\partial z_k}\right)_{i=1,2}, \qquad S'_{ik}(\mathbf{w}) = \sum_{k=1}^{2}\left(J_{ki}\frac{\partial w_j}{\partial z_k} + J_{kj}\frac{\partial w_i}{\partial z_k}\right).$$

Let $(\mathbf{w}', p')$ be a solution to auxiliary problem (10.4.15) in Ω'. When we subtract relations (10.4.15) for $(\mathbf{w}', p')$ from (10.4.23), we arrive at the following problem for the differences $\mathbf{u} = \mathbf{w} - \mathbf{w}'$, $s = p - p'$:

$$\begin{aligned} &-\nu\nabla^2\mathbf{u} + \nabla s = \mathbf{f}, \qquad \nabla\cdot\mathbf{u} = \rho, \qquad z \in \Omega', \\ &\mathbf{u}|_{\Sigma} = 0, \qquad \mathbf{u}\cdot\mathbf{n}'|_{\Gamma'} = b, \qquad \tau'\cdot S(\mathbf{u})\mathbf{n}'|_{\Gamma'} = d, \\ &\int_{-R_0}^{R_0} u_2\,dx_1 = 0, \qquad x_2 < -1. \end{aligned}$$

Here

$$\begin{aligned} \mathbf{f} &= \nu(\nabla'^2 - \nabla^2)\mathbf{w} + (\mathbf{a}\cdot\nabla)\mathbf{u} + (\mathbf{a}\cdot\nabla' - \nabla)\mathbf{w} \\ &\quad - (\mathbf{w}\cdot\nabla')\mathbf{w} + (\mathbf{w}'\cdot\nabla)\mathbf{w}' - (\nabla' - \nabla)p, \\ \rho &= -(\nabla' - \nabla)\cdot\mathbf{w}, \\ b &= -\mathbf{w}\cdot(J^* - I)\mathbf{n}' + \mathbf{a}(J^* - I)\mathbf{n}', \\ d &= -(\hat{J} - I)\tau'\cdot S'(\mathbf{w})J^*\mathbf{n}' - \tau'\cdot[S'(\mathbf{w}) - S(\mathbf{w})]J^*\mathbf{n}' - \tau'\cdot S(\mathbf{w})(J^* - I)\mathbf{n}'. \end{aligned}$$

Clearly, the supports of $\mathbf{f}$ and ρ are contained in the domain $\{x \in \Omega :\ x_2 \ge -1\}$. Making use of inequalities (10.4.10) and (10.4.22), we obtain

$$|\mathbf{u}|_{\overset{o}{C}{}^{2+\alpha}_{\beta}(\Omega)} + |s|_{\overset{o}{C}{}^{1+\alpha}_{\beta-1}(\Omega)} \le c_{13}a|r - r'|_{\overset{o}{C}{}^{3+\alpha}_{1+\beta}(I)}$$

and, as a consequence,

$$|t[r] - t[r']|_{\overset{o}{C}{}^{1+\alpha}_{\beta-1}(I)} \le c_{14}a|r - r'|_{\overset{o}{C}{}^{3+\alpha}_{1+\beta}(I)}. \tag{10.4.24}$$

Hence, for r, r' from the ball $\mathcal{B}_\delta$ we have

$$|\mathcal{M}r|_{\overset{o}{C}{}^{3+\alpha}_{1+\beta}(I)} \leq c_{15}\delta^2 + c_{16}a,$$
$$|\mathcal{M}r - \mathcal{M}r'|_{\overset{o}{C}{}^{3+\alpha}_{1+\beta}(I)} \leq c_{17}(\delta + a)|r - r'|_{\overset{o}{C}{}^{3+\alpha}_{1+\beta}(I)}.$$

This shows that $\mathcal{M}$ is a contraction operator in the above ball, if δ is such that

$$c_{15}\delta^2 + c_{16}a \leq \delta, \qquad c_{17}(\delta + a) < 1.$$

Clearly, these inequalities can be satisfied if δ and a are small. Under these conditions, equation (10.4.20) is uniquely solvable in $\mathcal{B}_\delta$.

It remains to construct the transformation Y. Let Ω_0 be a domain corresponding to the rest state, i.e., the domain with the boundary $\partial\Omega_0 = \Gamma_0 \cup \Sigma$ where Γ_0 is the semicircle $|x| = R_0$, $x_2 > 0$. We construct the transformation $Y_0 : \Omega_0 \to \Omega$ with the properties (1)–(3) in the form

$$x_i = z_i + \Phi_i(z), \qquad i = 1, 2, \quad z \in \Omega_0.$$

Functions $\Phi_i(z)$ vanish on Σ, $\Phi_i = (R - R_0)z_i|z|^{-1}$ on Γ_0, and Φ_i are extended into Ω_0 with the preservation of class, i.e., $\Phi_i \in \overset{o}{C}{}^{3+\alpha}_{1+\beta}(\Omega_0)$ and

$$|\Phi_i|_{\overset{o}{C}{}^{3+\alpha}_{1+\beta}(\Omega_0)} \leq c_{18}|R - R_0|_{\overset{o}{C}{}^{3+\alpha}_{1+\beta}(I)}.$$

The construction of such an extension is given in [8], for example. It can be carried out in such a way that $\Phi_1 \in \overset{o}{C}{}^{3+\alpha}_{1+\beta,-}(\Omega_0)$, $\Phi_2 \in \overset{o}{C}{}^{3+\alpha}_{1+\beta,+}(\Omega_0)$, and $\Phi_i(z) = 0$ for $z_2 < 0$ (the latter property may be obtained after multiplication of the extended functions by appropriate cut-off functions, see (10.1.6)). Clearly, the mapping Y_0 is invertible for small δ. Moreover, since the extension operator is linear, for the arbitrary two functions R, R' such that $R - R_0$, $R' - R_0 \in \mathcal{B}_\delta$ the corresponding Φ_i and Φ_i' satisfy the inequality

$$|\Phi_i - \Phi_i'|_{\overset{o}{C}{}^{3+\alpha}_{1+\beta}(\Omega_0)} \leq c_{18}|R - R'|_{\overset{o}{C}{}^{3+\alpha}_{1+\beta}(I)},$$

and for the Jacobi matrices of the transformations Y_0 and Y_0' the estimate

$$|J_0 - J_0'|_{\overset{o}{C}{}^{2+\alpha}_{\beta}(\Omega_0)} \leq c_{19}|R - R'|_{\overset{o}{C}{}^{3+\alpha}_{1+\beta}(I)} \tag{10.4.25}$$

holds. The mapping $Y : \Omega \to \Omega'$ mentioned above may be defined as follows: $Y = Y_0' \circ Y_0^{-1}$. Then $J - I = J_0^{-1}(J_0' - J_0)$, and inequality (10.4.22) follows from (10.4.25).

The proof of Theorem 10.4.1 is now complete. ■

We observe at the conclusion that almost all the arguments presented in this section apply in the case $a < 0$. The positivity of a was used in the previous sections 10.2 and 10.3.

10.5. The Piston Problem

In this section we consider a variant of the free boundary problem treated in Section 10.4. The domain Ω occupied by the liquid is bounded and $\partial\Omega = \Sigma_1 \cup \Sigma_2 \cup \Gamma$ where

$$\Sigma_1 = \{|x_1| \leq R_0,\ x_2 = 0\}, \qquad \Sigma_2 = \{|x_1| = \pm R_0,\ 0 < x_1 < h\},$$

and Γ is a curve with endpoints $x_\pm = (\pm R_0, h)$ contained in the strip $|x_1| < R_0$. It is required to find $\mathbf{w}(x) = (w_1, w_2),\ p(x)$, and the free boundary Γ satisfying the relations

$$\begin{aligned} -\nu\nabla^2\mathbf{w} - (\mathbf{a}\cdot\nabla)\mathbf{w} + (\mathbf{w}\cdot\nabla)\mathbf{w} + \nabla p = 0, \qquad \nabla\cdot\mathbf{w} = 0, \qquad x \in \Omega, \\ \mathbf{w}|_{\Sigma_1} = \mathbf{a} = (0, a), \qquad a > 0, \quad \mathbf{w}|_{\Sigma_2} = 0, \qquad (10.5.1) \\ \sigma H\mathbf{n} - T(\mathbf{w}, p)\mathbf{n}|_\Gamma = 0, \qquad \mathbf{w}\cdot\mathbf{n} - an_2|_\Gamma = 0. \end{aligned}$$

In addition, we assume that the contact angle equals π and we prescribe the volume of the liquid:

$$|\Omega| = Q. \qquad (10.5.2)$$

The ordinate h of the contact points is not prescribed.

The line Σ_1 can be interpreted as a bottom moving uniformly with the velocity a ("a piston"). The free boundary problem (10.5.1), (10.5.2) is written in a coordinate system moving together with the piston.

The piston problem was considered by D. Kröner [4] under the assumption that on both Σ_1 and Σ_2 a partial slip condition holds

$$\begin{aligned} \mathbf{w}\cdot\mathbf{n}|_{\Sigma_1\cup\Sigma_2} = 0, \qquad \nu\tau\cdot S(\mathbf{w})\mathbf{n} + \gamma\mathbf{w}\cdot\tau|_{\Sigma_2} = 0, \\ \nu\tau\cdot S(\mathbf{w})\mathbf{n} + \gamma(\mathbf{w} - \mathbf{a})\cdot\tau|_{\Sigma_1} = 0, \end{aligned}$$

and that the position of the contact points $x_\pm$ is prescribed, whereas the contact angle is found in the process of the solution of the problem.

Except for the points $x_\pm$, in this problem there are two more singular points $y_\pm = (\pm R_0, 0)$ where the piston meets the walls $|x_1| = R_0$ of the container. In these points the velocity vector field has a discontinuity jump and, as a consequence, the Dirichlet integral of $\mathbf{w}$ is unbounded in their neighborhoods. We will find the solution of this problem in a special weighted space whose elements are bounded in $\bar{\Omega}$, smooth in the interior of Ω but they need not be continuous at $y_\pm$. Let $\rho_{s,k}(x)$ be smooth functions which are positive in $\bar{\Omega}$ except probably the singular points $x_\pm$, $y_\pm$, and

$$\rho_{s,k}(x) = \begin{cases} |x - x_\pm|^{k-s} & \text{in the neighborhoods of } x_\pm, \\ |x - y_\pm|^k & \text{in the neighborhoods of } y_\pm, \end{cases}$$
$$(k \geq 0,\ s \in \mathbb{R}).$$

By $\overset{\circ}{C}{}^{l}_{s,0}(\Omega)$ we mean the space of functions which are C^l-smooth in $\bar\Omega \setminus M$, $M = \{x_+ \cup x_- \cup y_+ \cup y_-\}$, C^s-smooth in $\bar\Omega \setminus \{y_+ \cup y_-\}$, bounded in $\bar\Omega$, and have a finite norm

$$|u|_{\overset{\circ}{C}{}^{l}_{s,0}(\Omega)} = \sum_{|j|=k<l} \sup_{x\in\Omega} |D^j u(x)|\rho_{s,k}(x)$$
$$+ \sum_{|j|=l} \sup_{x\in\Omega} \rho_{s,l}(x) \sup_{\substack{y\in\Omega \\ |x-y|\le\frac{1}{2}d(x)}} \frac{|D^j u(x) - D^j u(y_|}{|x-y|^{l-[l]}}, \tag{10.5.3}$$

where $d(x) = \operatorname{dist}(x, M)$. We can also define spaces $\overset{\circ}{C}{}^{l}_{s,t}(\Omega)$ with the norm (10.5.3) where $\rho_{s,k}(x)$ is replaced by the function $\rho_{s,t,k}(x)$ such that

$$\rho_{s,k,t}(x) = \begin{cases} |x - x_\pm|^{k-s} & \text{in the neighborhoods of } x_\pm, \\ |x - y_\pm|^{k-t} & \text{in the neighborhoods of } y_\pm. \end{cases}$$

For $s, t > 0$, elements of $\overset{\circ}{C}{}^{l}_{s,t}(\Omega)$ are Hölder continuous in $\bar\Omega$.

Our main result in this section is as follows.

Theorem 10.5.1. *If $a > 0$ is sufficiently small and*

$$|\Omega| = Q > \frac{1}{2}\pi R_0^2, \tag{10.5.4}$$

then problem (10.5.1), (10.5.2) has a unique solution with the following properties:

(1) $\mathbf{w} \in \overset{\circ}{C}{}^{2+\alpha}_{\beta,0}(\Omega)$, $p \in \overset{\circ}{C}{}^{1+\alpha}_{\beta-1,-1}(\Omega)$ *with* $\alpha \in (0,1)$, $\beta \in (0, \frac{1}{2})$;

(2) *The curve Γ is located in the strip $|x_1| \le R_0$, and it is given by the equation of type (10.4.6), i.e.,*

$$r = |x - x_h| = R(\varphi) \tag{10.5.5}$$

where (r, φ) are standard polar coordinates with the center in $x_h = (0, h)$, $h = (x_\pm)_2$, and $R - R_0 \in \overset{\circ}{C}{}^{3+\alpha}_{1+\gamma,+}(I)$ with $\gamma \in (\frac{1}{2}, 1)$.

(3) *$w_2(x)$, $p(x)$ are even and $w_1(x)$ is an odd function of x_1;*

(4) *There holds the estimate*

$$|\mathbf{w}|_{\overset{\circ}{C}{}^{2+\alpha}_{\beta,0}(\Omega)} + |\nabla p|_{\overset{\circ}{C}{}^{\alpha}_{\beta-2,-2}(\Omega)} + |R - R_0|_{\overset{\circ}{C}{}^{3+\alpha}_{1+\gamma}(I)} \le c_1 a.$$

The proof of this theorem is based on some auxiliary propositions. We begin with the determination of the main term of the asymptotics of the solution near $y_\pm$.

Theorem 10.5.2. *Consider the Stokes problem*

$$-\nu\nabla^2\mathbf{v} + \nabla p = 0, \qquad \nabla \cdot \mathbf{v} = 0, \qquad x \in d_\vartheta,$$
$$\mathbf{v}|_{\gamma_\vartheta} = 0, \qquad \mathbf{v}|_{\gamma_0} = \mathbf{b} = \text{const}, \tag{10.5.6}$$

where d_ϑ is an infinite plane wedge of the opening $\vartheta \in (0, 2\pi)$ given by the relations $0 < \varphi < \vartheta$, $\rho > 0$, ρ, φ are polar coordinates on the plane $\mathbb{R}^2$ and $\gamma_0, \gamma_\vartheta$ are its sides $\varphi = 0$ and $\varphi = \vartheta$, respectively. The solution of problem (10.5.6) *has the form*

$$\mathbf{v} = \mathbf{V}(\varphi), \qquad p = \rho^{-1}q(\varphi), \tag{10.5.7}$$

$$V_1 = b_1 + A_1 \sin^2\varphi - A_2(\varphi + \sin\varphi\cos\varphi)$$
$$V_2 = b_2 + A_1(\varphi - \cos\varphi\sin\varphi) - A_2\sin^2\varphi,$$
$$q = -2\nu(A_1\cos\varphi + A_2\sin\varphi), \tag{10.5.8}$$

where

$$A_1 = \frac{b_1\sin^2\vartheta - b_2(\vartheta + \sin\vartheta\cos\vartheta)}{\vartheta^2 - \sin^2\vartheta}, \qquad A_2 = \frac{b_1(\vartheta - \cos\vartheta\sin\vartheta) - b_2\sin^2\vartheta}{\vartheta^2 - \sin^2\vartheta}.$$

Proof. We find at first the solution of the Stokes equation with boundary conditions

$$\mathbf{v}|_{\gamma_\vartheta} = \mathbf{b}, \qquad \mathbf{v}|_{\gamma_0} = 0, \tag{10.5.9}$$

which also has form (10.5.7), i.e., $\mathbf{v} = \mathbf{V}'(\varphi)$, $p = \rho^{-1}q'(\varphi)$. The radial and angular components v' and w' of $\mathbf{V}'$ should satisfy (together with q') system (10.2.2) with $\lambda = 0$ whose linearly independent solutions are columns of the matrix

$$\begin{pmatrix} -\varphi\sin\varphi & \varphi\cos\varphi + \sin\varphi & \cos\varphi & \sin\varphi \\ \sin\varphi - \varphi\cos\varphi & -\varphi\sin\varphi & -\sin\varphi & \cos\varphi \\ 2\nu\cos\varphi & 2\nu\sin\varphi & 0 & 0 \end{pmatrix}.$$

Because of the boundary conditions (10.5.9), (v', w', q') are linear combinations of the first two columns with the coefficients A_1 and A_2 determined from the system

$$-A_1\vartheta\sin\vartheta + A_2(\vartheta\cos\vartheta + \sin\vartheta) = b',$$
$$A_1(\sin\vartheta - \vartheta\cos\vartheta) - A_2\vartheta\sin\vartheta = b'',$$

where $b' = b_1\cos\vartheta + b_2\sin\vartheta$ and $b'' = -b_1\sin\vartheta + b_2\cos\vartheta$ are radial and angular components of $\mathbf{b}$ on γ_ϑ, respectively. We see that

$$A_1 = -\frac{b'\vartheta\sin\vartheta + b''(\vartheta\cos\vartheta + \sin\vartheta)}{\vartheta^2 - \sin^2\vartheta} = \frac{b_1\sin^2\vartheta - b_2(\vartheta + \sin\vartheta\cos\vartheta)}{\vartheta^2 - \sin^2\vartheta},$$
$$A_2 = -\frac{b'(\sin\vartheta - \vartheta\cos\vartheta) + b''\vartheta\sin\vartheta}{\vartheta^2 - \sin^2\vartheta} = \frac{b_1(\vartheta - \sin\vartheta\cos\vartheta) - b_2\sin^2\vartheta}{\vartheta^2 - \sin^2\vartheta}.$$

The solution of problem (10.5.6) is $\mathbf{v}(\varphi) = \mathbf{b} - \mathbf{V}''(\varphi) = \mathbf{V},\ p = \rho^{-1}q'(\varphi)$ and is given by (10.5.8). The theorem is proved. ■

It is easy to see that $v_1 = -\frac{\partial\phi}{\partial x_2},\ v_2 = \frac{\partial\phi}{\partial x_1}$ where

$$\phi = \rho(b_2\cos\varphi - b_1\sin\varphi) + \rho[A_2\varphi\sin\varphi - A_1(\sin\varphi - \varphi\cos\varphi)].$$

In particular, when $\vartheta = \pi/2,\ \mathbf{b} = (0, a)$, we have

$$\begin{aligned} A_1 &= -\frac{2a\pi}{\pi^2-4}, \qquad A_2 = -\frac{4a}{\pi^2-4}, \\ V_1 &= -\frac{2a\pi}{\pi^2-4}\sin^2\varphi + \frac{4a}{\pi^2-4}(\varphi + \cos\varphi\sin\varphi), \\ V_2 &= a - \frac{2a\pi}{\pi^2-4}(\varphi - \cos\varphi\sin\varphi) + \frac{4a}{\pi^2-4}\sin^2\varphi, \\ p &= \frac{2\nu}{\rho}\left(\frac{2a\pi}{\pi^2-4}\cos\varphi + \frac{4a}{\pi^2-4}\sin\varphi\right), \\ \phi &= \rho a\cos\varphi + \frac{2a\rho}{\pi^2-4}[-2\varphi\sin\varphi + \pi(\sin\varphi - \varphi\cos\varphi)]. \end{aligned} \tag{10.5.10}$$

Theorem 10.5.3. *Let Ω be a domain bounded by $\Sigma_1,\ \Sigma_2$ and by the curve Γ determined by equation* (10.5.5) *with a given $R(\varphi)$ such that $R - R_0 \in \overset{\circ}{C}{}^{3+\alpha}_{1+\beta,+}(I)$. If a is small enough, then the problem*

$$\begin{aligned} -\nu\nabla^2\mathbf{w} - (\mathbf{a}\cdot\nabla)\mathbf{w} + (\mathbf{w}\cdot\nabla)\mathbf{w} + \nabla p = 0, \qquad \nabla\cdot\mathbf{w} = 0, \qquad x \in \Omega, \\ \mathbf{w}|_{\Sigma_1} = \mathbf{a}, \qquad \mathbf{w}|_{\Sigma_2} = 0, \\ \mathbf{w}\cdot\mathbf{n} - an_2|_\Gamma = 0, \qquad \tau\cdot S(\mathbf{w})\mathbf{n}|_\Gamma = 0, \end{aligned} \tag{10.5.11}$$

has a unique solution $\mathbf{w} \in \overset{\circ}{C}{}^{2+\alpha}_{\beta,0}(\Omega),\ p \in \overset{\circ}{C}{}^{1+\alpha}_{\beta-1,-1}(\Omega)$ satisfying condition (10.4.2). *It possesses the symmetry properties indicated in Theorem* 10.5.1 *and satisfies the inequality*

$$|\mathbf{w}|_{\overset{\circ}{C}{}^{2+\alpha}_{\beta,0}(\Omega)} + |p|_{\overset{\circ}{C}{}^{1+\alpha}_{\beta-1,-1}(\Omega)} \le c_2 a. \tag{10.5.12}$$

Proof. We introduce the auxiliary solenoidal vector field

$$\mathbf{w}_0 = \left(-\frac{\partial\phi\chi(x)}{\partial x_2}, \frac{\partial\phi\chi(x)}{\partial x_1}\right)$$

satisfying boundary conditions (10.5.11) near the points $y_\pm$. Here $\chi(x)$ is a smooth function equal to 1 for $|x-y_+| \le d/2,\ |x-y_-| \le d/2$ and to zero for $|x-y_\pm| \ge d$ with $d < \max(h, R_0/2)$, and ϕ is the stream function defined by (10.5.10) near $y_\pm$ (polar coordinates are chosen in the neighborhoods of $y_\pm$ in such a way that Σ_1 corresponds to $\varphi = 0$ and Σ_2 corresponds to $\varphi = \pi/2$). We may suppose that $\chi = \chi(r)$. We also introduce the function $p_0 = \rho^{-1}q(\varphi)\chi$ where q is given by (10.5.8) near $y_\pm$. For $\mathbf{u} = \mathbf{w} - \mathbf{w}_0,\ q = p - p_0$ we have

$$\begin{gathered}-\nu\nabla^2\mathbf{u}+(\mathbf{w}_0-\mathbf{a}\cdot\nabla)\mathbf{u}+(\mathbf{u}\cdot\nabla)\mathbf{w}_0+(\mathbf{u}\cdot\nabla)\mathbf{u}+\nabla q=\mathbf{f}-(\mathbf{w}_0-\mathbf{a}\cdot\nabla)\mathbf{w}_0,\\ \nabla\cdot\mathbf{u}=0,\qquad \mathbf{u}|_{\Sigma_1}=\mathbf{a}-\mathbf{w}_0\equiv\alpha,\qquad \mathbf{u}|_{\Sigma_2}=0, \qquad (10.5.13)\\ \mathbf{u}\cdot\mathbf{n}-an_2|_\Gamma=0,\qquad \tau\cdot S(\mathbf{u})\mathbf{n}|_\Gamma=0,\end{gathered}$$

where $\mathbf{f}=\nu\nabla^2\mathbf{w}_0-\nabla p_0$ is a smooth function whose support is contained in the set $d/2<|x-y_\pm|<d$. Since $\operatorname{supp}\alpha\in\{-R_0+d/2<x_1<R_0-d/2,\ x_2=0\}$, we can easily prove that problem (10.5.13) has a solution $\mathbf{u}\in\overset{o}{C}{}^{2+\alpha}_{\beta,\delta}(\Omega)$, $\nabla q\in\overset{o}{C}{}^{\alpha}_{\beta-2,\delta-2}(\Omega)$ with some $\delta\in(0,1)$. Clearly, $\mathbf{w}=\mathbf{u}+\mathbf{w}_0$, $p=q+p_0$ is a solution of (10.5.11). The uniqueness of this solution follows from the results of paper [15], since for every $r\in(0,d)$ we have

$$\int_{\Omega(r)}|\nabla\mathbf{w}|^2\,dx\le c_3a\ln\frac{d}{r},$$

where $\Omega(r)=\{x\in\Omega:\operatorname{dist}|x-y_\pm|>r\}$. Inequality (10.5.12) follows from the definition of $\mathbf{w}_0$, p_0, and from the estimate

$$\begin{aligned}&|\mathbf{u}|_{\overset{o}{C}{}^{2+\alpha}_{\beta,\delta}(\Omega)}+|\nabla q|_{\overset{o}{C}{}^{\alpha}_{\beta-2,\delta-2}(\Omega)}\\ &\quad\le c_4\big(|\mathbf{f}-(\mathbf{w}_0-\mathbf{a}\cdot\nabla)\mathbf{w}_0|_{\overset{o}{C}{}^{\alpha}_{\beta-2,\delta-2}(\Omega)}+|\alpha|_{\overset{o}{C}{}^{2+\alpha}_{\delta}(\Sigma_1)}+|an_2|_{\overset{o}{C}{}^{2+\alpha}_{\beta}(\Gamma)}\big)\le c_5a\end{aligned}$$

for the solution of (10.5.13). The theorem is proved. ∎

Proof of Theorem 10.5.1. If $a=0$, then $\mathbf{v}=0$, $p=-F_0=\sigma R_0^{-1}$, and Γ is the semicircle $x_1^2+(x_2-h_0)^2=R_0^2$, $x_2>h_0$. The value of h_0 is found from the relation

$$Q=2R_0h_0+\tfrac{1}{2}\pi R_0^2.$$

Clearly, $h_0>0$ by virtue of (10.5.4). For small $a>0$, we may represent the pressure in the form

$$p=P-F,$$

and if $(\mathbf{w},P)$ satisfies condition (10.4.2), then $F=F_0$. As in Section 10.4, the whole free boundary problem can be reduced to the equation on Γ:

$$\sigma H-\mathbf{n}\cdot T(\mathbf{w},P)\mathbf{n}|_\Gamma=F, \qquad (10.5.14)$$

where $(\mathbf{w},P)$ is a solution of the auxiliary problem (10.5.11).

Further arguments are almost exactly the same as in Theorem 10.4.1. It is convenient to assume that the contact points have coordinates $(\pm R_0,0)$ and

$$\Sigma_1=\{|x_1|\le R_0,\ x_2=-h\},\qquad \Sigma_2=\{x_1=\pm R_0,-h<x_1<0\}. \quad (10.5.15)$$

Equation (10.5.14) can be written in the form (10.4.17) and (10.4.20) with $r=R-R_0$. The function $t[r]=\mathbf{n}\cdot T(\mathbf{w},P)\mathbf{n}$ satisfies estimates (10.4.21), (10.4.24) which can be proved by the same arguments as in Section 10.4, only the mapping Y should be slightly modified. Let Ω and Ω' be two domains with the same area: $|\Omega| = |\Omega'| = Q$ bounded by the contours $\Sigma_1 \cup \Sigma_2 \cup \Gamma$ and $\Sigma_1' \cup \Sigma_2' \cup \Gamma'$, respectively, where Σ_i and Σ_i' are defined by (10.5.15) and by

$$\Sigma_1' = \{|x_1| < R_0,\ x_2 = -h'\}, \qquad \Sigma_2' = \{x_1 = \pm R_0,\ -h' < x_1 < 0\}, \qquad h' > 0,$$

and Γ, Γ' are given by the equations

$$|x| = R(\varphi) = R_0 + r(\varphi), \qquad |x| = R'(\varphi) = R_0 + r'(\varphi),$$

respectively. We suppose that $r, r' \in \overset{\circ}{C}{}^{3+\alpha}_{1+\beta,+}(I)$ but the curves Γ and Γ' need not be contained in the strip $|x_1| \leq R_0$. Clearly,

$$Q = 2R_0 h + \frac{1}{2}\int_0^\pi (R_0 + r)^2\, d\varphi = 2R_0 h' + \frac{1}{2}\int_0^\pi (R_0 + r')^2\, d\varphi,$$

hence,

$$h - h' = \frac{1}{4R_0}\int_0^\pi (2R_0 + r + r')(r' - r)\, d\varphi.$$

The mapping $Y : \Omega \to \Omega'$ with all the necessary properties can be defined as $Y = Y_2 \circ Y_1$ where Y_1 is the transformation from Theorem 10.4.1 and $Y_2 x = y$ is defined by

$$y_1 = x_1, \qquad y_2 = x_2 + (h' - h)\tau(x_2),$$

where $\tau(x_2)$ is a smooth function equal to 1 for $x_2 \leq -\min(h, h')$ and vanishing for $x_2 \geq 0$.

Inequality (10.4.24) is easily verified. All the subsequent arguments are exactly the same as in Theorem 10.4.1. Theorem 10.5.1 is proved. ■

References

[1.] Pukhnachov, V.V., and Solonnikov, V.A. On the problem of dynamic contact angle. *Prikl. Mat. Mekh. USSR*, **46** (1983), 771–779.

[2.] Pukhnachov, V.V. Plane steady-state free boundary problem for the Navier–Stokes equations. *Zh. Prikl. Mekh. Tekhn. Phys.*, **3** (1972), 91–102.

[3.] Solonnikov, V.A. Solvability of the problem on the plane motion of a heavy viscous incompressible capillary liquid partially filling a container. *Math. USSR Izv.*, **14** (1980), 193–221.

[4.] Kröner, D. Asymptotische Entwicklungen für Strömungen von Flüssigkeiten mit freiem Rand und dynamischem Kontaktwinkel. Preprint, no. 809, Bonn, 1986, pp. 1–105. See also: Lecture Notes in Mathematics, vol **143**. Springer-Verlag, New York, 1990, pp. 49–59.

[5.] Baiocchi, C., and Pukhnachov, V.V. Problems with unilateral restrictions for the Navier–Stokes equations and the problem of dynamic contact angle. *Zh. Prikl. Mekh. Tehkn. Phys.*, **2** (1990), 27–40.

[6.] Sokolowsky, J. On a free boundary problem for the stationary Navier–Stokes equations with a dynamic contact line. In *The Navier–Stokes Equations II. Theory and Numerical Methods* (J. Heywood, K. Masuda, R. Rautmann, and V. Solonnikov, eds.). Lecture Notes in Mathematics, vol. **1530**, Springer-Verlag, New York, 1992, pp. 17–29.

[7.] Antanovskii, L.K. Boundary integral equations for contact problems of plane quasi-steady viscous flow. *Europ. Journ. of Appl. Math.*, **4**, 2, (1993), 175–188.

[8.] Solonnikov, V.A. On the Stokes equations in domains with non-smooth boundaries and on viscous flow with a free surface. In *Nonlinear Partial Differential Equations and Their Applications*. College de France Seminars, vol. **III** (H. Brezis and J.-L. Lions, eds.), 1981, 340–423.

[9.] Kondratiev, V.A. Boundary value problems for elliptic equations in domains with conical and angular points. *Trans. Moscow Math. Soc.*, **16** (1967), 209–292.

[10.] Maz'ya, V.G., and Plamenevskii, B.A. Estimates in L_p and in Hölder spaces and the Miranda–Agmon maximum principle for solutions of elliptic boundary value problems with singular points at the boundary. *Math. Nachr.* **81** (1978), 25–82.

[11.] Maz'ya, V.G., and Plamenevskii, B.A. On the coefficients in the asymptotics of solutions of elliptic boundary value problems in domains with conical points. *Math. Nachr.* **76** (1977), 29–60.

[12.] Ladyzhenskaya, O.A., and Solonnikov, V.A. Determination of solutions of boundary value problems for steady-state Stokes and Navier–Stokes equations having an infinite Dirichlet integral. *J. Soviet Math.*, **21** (1983), 728–761.

[13.] Pileckas, K. Solvability of a problem on a plane motion of a viscous incompressible liquid with a free non-compact boundary. *Differentsial'nye Uravneniya i Primenen*, **30** (1981), 57–95.

[14.] Solonnikov, V.A. Solvability of the problem of effluence of a viscous incompressible fluid into an infinite open basin. *Proc. Steklov Inst. Math.*, no. 2 (1989), 193–225.

[15.] Solonnikov, V.A. On boundary value problems for the Navier–Stokes equations with discontinuous boundary data. *Partial Differential Equations and Related Subjects*. Proc. Conf. in Trento, Sept. 1990, M. Miranda (editor). Pitman Research Notes in Mathematics Series, **269** (1992), pp. 227–237.

11
Space, Time and Energy

James Serrin

A not inaccurate paraphrase of modern cosmological opinion is that our Universe initiated in, and expanded from, an incendiary moment some twenty billion years ago, and is governed by the resulting space–time structure and the induced curvature of space, as seen through the eyes of Albert Einstein. Moreover, it is held that Isaac Newton's classical world view—though complete and perfect, arguably the pinnacle of man's intellectual achievement—nevertheless does not meet the requirements of modern cosmology and can confidently be discarded as a guide.

These views are persuasively presented in numberless popular expositions of science, a recent and notable example being the best-selling book *A Brief History of Time*; its catchy and marvelously paradoxical title says everything: it is an invitation to the painless acquisition of wisdom. I would, however, advance a contrary case, that the last cosmological word is not in, and may never be delivered to us. If scholarship and ability are not snuffed out by ideology and conformity, then the mythologies of our time should have no better chance of survival than those of earlier days.

In a letter to his patron Chantelou, the artist Nicholas Poussin wrote in 1642:

> Things which partake of perfection should not be looked at in haste, but call for time, judgment, and intelligence. The means employed in their appraisal must be the same as those used in their making.

These are remarkable words, their originality no doubt even more striking today than when they were first written. They supply the motif for this lecture, my thesis being that we can indeed look with deeper understanding at Newtonian mechanics, that a sharper perception of its structure emerges when we take into account its underlying relation with the geometry of space and time.

I suppose no apology need be given for engaging your interest in such classical matters, especially in a university where Copernicus and Galileo are honored. The science of our time, just as the science of past centuries, depends crucially and precisely on the mechanics of Newton. It forms the basis for our technological society, for the curriculum in every engineering school in the world, for the successful exploration of space. Dare I say that every gain which we achieve in understanding the Newtonian world should be reflected later in the understanding of the vaster cosmological universe.

Newton's ideas and his life work, the *Philosophiae Naturalis Principia Mathematica*, originally captured the public imagination with its revelation, explanation, and prediction of the phenomena of celestial mechanics. Our science and our world has not since been the same. At the heart of the issue is the power that arises from the interrelation between acceleration, as a second derivative of position, and the notion of universal forces.

I have introduced the three threads of the lectures: space, time, and energy. How are they to be tied together, placed in a mathematical context so that their interrelation can teach us something new about the world? Here we tread on difficult, though hallowed ground. The great physicists of the past already treated these granitic concepts in their struggle to understand the physical world: Newton, and before him Galileo; Leibniz; Sadi Carnot, Joule, Lord Kelvin and Rudolph Clausius—the founders of thermodynamics; George Gabriel Stokes—who saw at least dimly the central role of geometry in mechanics; Leonhard Euler and Augustin Louis Cauchy, who found the key ideas of internally organized forces.

The public barely knows these pioneers of physical science, yet they are the peers of Palestrina, Bach, Mozart, Handel, and Verdi in the musical world. The musical analogy is apt. What we may add to the original discoveries of natural philosophy cannot vie with the original conceptions, but as Berlioz, Brahms, and Respighi contributed still to music's power to move us, so we may hope to supply something new as well.

Returning to the threads of space, time, and energy, let me explain as simply as I can, how these threads are already woven into mechanics. For a moment, let us go back to Galileo. Galileo lived, as he thought, in a Euclidean world; he saw time as a uniform stream divided only by the markings of simple clocks, and space a Euclidean. We would say that his world is a classical space–time manifold of four dimensions, having associated to each instant of time a three-dimensional Euclidean spatial simultaneity sheet.

How do these simultaneity sheets join together as time moves forward? That is, if we imagine that each observer of the world sees himself in some preferred Cartesian frame in the manifold, and watches events transpire in that frame, how are the frames (charts) of different observers related? For Galileo they simply stand in uniform motion; in any simultaneity sheet they move relative to each other, with a uniform and fixed translational velocity.

Galileo's structure cannot, however, satisfy the needs of Newtonian mechanics, nor the philosophical penetration of Liebniz. For them, relatively rotating and moving frames are surely required—the rotating Earth we live on, or, for that matter, satellite technology, immediately forces this issue on us. More fundamentally, Newton's work leads to a coherent unification of mechanical and inertial effects, and externally operating forces. For this purpose, Newton had a felt need for one ultimately preferred frame, the absolute fixed space of God, an unmoving so-called inertial frame set into the distant fixed stars and in which all forces arise by material interaction. Leibniz saw the hand of God differently—that each frame, in the family of all possible relatively rotating and moving frames, was the equal of any other. Thus, in accordance with his doctrine of monads, there could not be individually preferred frames, and a fortiori no special inertial frame. The idea of the "fixed stars" was simply not relevant.

Historically, Leibniz lost the battle, partly through a change in philosophical focus, partly because mechanics became ever more the domain of the scientist rather than the natural philosopher. There is more to say, but first let me turn to the third thread, energy.

This is woven into the tapestry of mechanics through the First Law of Thermodynamics. Let us imagine a given material object, or physical system, which we shall call S, and focus our interest on it. Associated to S is its *total energy*—whatever else, a scalar time-dependent attribute of S, calculated in terms of the state variables which delineate S. That it is not an easy idea should be manifest from the fact that the concept did not occur in science for almost 200 years following Newton's initial discoveries. The physical chemist Peter Atkins has commented that:

> The Second Law of Thermodynamics [is] in my opinion much easier to understand than the [law of] conservation of energy. Energy is a word so familiar to us today that we can hardly grasp either the intellectual Everest it represents or the conceptual difficulty we face in saying exactly what it is.

We may recall C.P. Snow's famous vignette, making the point that educated people should be able to discuss the Second Law of Thermodynamics as intelligently as they would a work of Shakespeare. While I have grave doubts that what passes for an education today would equip anyone to discuss sensibly either the Second Law or Shakespeare, it does seem that Lord Snow might have added the concept of energy to his list.

The First Law grants the primacy of energy, and states that the time rate of change of the total energy of a system is caused by, or more technically, balanced by the *rate of working* of exterior forces on a system together with the *rate of heat addition* to the system, i.e., in my only equation,

$$dt/dE = W + Q.$$

This relation, originally discovered in the mid-nineteenth century for quite restricted circumstances, was given broader application by C. Neumann in 1894, and finally codified in the encyclopedic and celebrated treatise on the Mechanics of Continua by Clifford Truesdell and Richard Toupin. Along with the momentum balance equations of Newton, Euler, and Cauchy, it constitutes the heart of classical mechanics.

It is well known to every student of natural philosophy, though not particularly emphasized in textbooks, that this structure exhibits a remarkable degree of internal consistency the more closely it is examined, an issue which has fascinated me ever since my earliest acquaintance with the subject.

The first indication that there may be an underlying structure which gives rise to this consistency came in a short and relatively unnoticed paper by Walter Noll in 1959, in which he showed that an abstract power balance implied an abstract version of the Cauchy equation (with the force and acceleration lumped together as a single quantity). Shortly afterward, this idea was extended by Albert Green and Ronald Rivlin, but then lay dormant until an important paper of Miroslav Šilhavý appeared in 1989. Šilhavý's main, and major, contribution was to show that *the total energy of a continuum system must split into an internal and a kinetic part*, the demonstration, moreover, being carried out without invoking Cauchy's motion equations. To see how remarkable this is, let me emphasize that in this approach the kinetic energy $|v|^2/2$ arises without any of the usual preparation from dynamics—it cannot be replaced by $|v|^4/4$, for example—not because this would give some unusual or new kind of dynamics, but rather because classical space–time forbids it! I can recall my astonishment, and disbelief, when Šilhavý first told me of this result, while we were walking together on a promenade in Valencia. The threads had come together at last, though I refused to believe it—until his paper appeared.

The explanation for Šilhavý's remarkable result, and indeed for Walter Noll's original idea, lies in the invariance of the energy balance equation in the Leibnizian group of frames, that is, the family of all relatively moving Cartesian charts on the simultaneity sheets. When one says that the energy balance equation is invariant, what is meant is that different observers occupying different frames, but each watching a given motion of the system S, each interpret the energy balance equation for the motion of S in their own frames and carry out *(in their frames)* the calculation for the energy, the work, and the heat according to *the same* strictly given formulas. The idea then is to transform each of these interpretations into a common frame, and then, *since all interpretations ultimately represent the same motion,* to glean from this fact further conditions which must be satisfied by the assigned structure. The surprising thing is the really great amount of information which can be obtained this way, by appropriately choosing relations between the frames. To these comments, one must add that the whole program owes its success also to the major contributions in the 1960s of Walter Noll and Bernard Coleman in their study of the effects of frame invariance on the constitutive equations of materials.

A main mathematical and physical difficulty in carrying out Šilhavý's program lies in the force structure which is allowed. Since we cannot make circular use of ultimate conclusions in order to obtain those same conclusions, there is no a priori reason why the forces in one frame should have any definite relation to those in another, beyond what is determined from the methodology of invariance. Indeed, even in the context of Newtonian mechanics, the transformation rules for forces from one frame to another are quite complicated, involving Coriolis and centrifugal effects, and others, even more subtle. From the Leibnizian point of view these forces retain a full reality, even as their form and their chart transformations must be a priori unknown.

To circumvent this difficulty requires a careful examination of the amount of freedom of reaction which a fixed force structure will allow to a physical system; the needed Freedom Axiom must simultaneously be strong enough to lead to Silhavý's Splitting Theorem, and at the same time one which is satisfied for ordinary dynamical systems.

Leaving aside these technically subtle matters, as we surely must for now, let me pass to further consequences of Šilhavý's theorem for the structure of mechanics. In fact, once it has been proved that the total energy splits into kinetic and internal parts, it is well known that the primary equations of momentum balance, the Cauchy equations themselves, are a consequence of, and are equivalent to, the group invariance of the energy balance equation. This is no trivial exercise: nonetheless, after the splitting theorem is proved there is a sufficient mathematical structure which we can exploit so that the equivalence becomes a straightforward matter.

In conclusion, in brief summary, the invariance of energy balance within the Leibnizian group of frames in classical space–time contains within it the *necessity* that the total energy of a physical system (of the type studied in continuum mechanics) must split into the sum of a kinetic energy and an internal energy; the coefficient of the squared velocity in the kinetic energy can be interpreted as the mass density of the system; and the Cauchy momentum balance equations are derived consequences of the structure.

Where does this leave the disagreement between Newton and Leibniz about the place of inertial frames in the foundations of mechanics? From what we have seen, certainly inertial frames are not *needed* for mechanics: everything in the above argument and in the final, striking conclusion that the Cauchy equations must hold because of space–time invariance, can be carried out without the intervention of an inertial frame. Invoking Occam's Razor as a philosophical tool, we will conclude in favor of Leibniz: what is not needed should not be presumed.

There is concealed in this summary an almost microscopic qualification, a parenthetical comment added almost in passing: the discussion applies to physical systems of the type studied in continuum mechanics and, it almost goes without saying, "to the systems of analytical dynamics." Is this qualification only a happenstance of the methodology? Are the Cauchy equations derivable from space, time, and energy considerations for every possible physical system? It would seem almost beyond conception that this could be true—the energy structures of general systems can be very intricate and indeed may even be unknowable—for instance, can we assign such a structure to the incredible complications of living things? But if this is the case, what then should we say about Newtonian mechanics as a general principle for nonrelativistic experience? I cannot dare to suggest an answer, but if Šilhavý's Splitting Theorem is not a universal attribute of materials then surely the world can turn out to be far more complicated even at the nonrelativistic level than we might presently imagine.

I have recently taken up with Professor Lawrence Markus the cosmologically complementary question of classifying four-dimensional space–time manifolds which are locally, though not globally, Leibnizian. This is motivated by the observation that if space–time is locally Leibnizian, then the local equations of motion still remain those of Newton, Cauchy, and Euler, that is, local mechanics is entirely unchanged. It is thus conceivable that what occurs at cosmological distances is influenced not only by mechanical principles but also by the fact that local Leibnizian charts may have no global extensions.

We are still left with an unexplained and inexplicable fact. If in practice the Solar System apparently floats in a genuine inertial frame, why is it that this inertial frame is anchored in the fixed stars, which after all are not really fixed in the first place? Moreover, the vanishing force structure which characterizes our local inertial frame may not have a global inertial extension. Indeed, if an inertial extension were not available, or if it did not exist, then there would arise the possibility that the cosmological data could be interpreted in alternative and different ways, depending on whatever actual extension might be chosen. Due to the high level of speculation already present in the subject, it might not be too much to add that even further speculation is supportable. It is certainly not my intention to expand further on these elusive ideas, but let me at least add that the nature of any postulated space–time structure, whether it be Newtonian or relativistic, could deeply influence cosmological speculation. For me, the hand of God continues to move mysteriously, and the paradigms of modern science are not yet impressed on the tablets of eternity. It is left to reiterate that the more closely one probes the classical mechanics of Newton, the more remarkable is the vista which opens to us; the questions which remain should all the more make us beware of mechanical and cosmological certitude.

[An earlier version of the lecture, in Italian, was published by the Università degli Studi di Ferrara in 1995.]

Index

MIX
Papier aus verantwortungsvollen Quellen
Paper from responsible sources
FSC® C105338

FSC
www.fsc.org

If you have any concerns about our products,
you can contact us on
ProductSafety@springernature.com

In case Publisher is established outside the EU,
the EU authorized representative is:
Springer Nature Customer Service Center GmbH
Europaplatz 3, 69115 Heidelberg, Germany

Printed by Libri Plureos GmbH
in Hamburg, Germany